# The Particle Garden

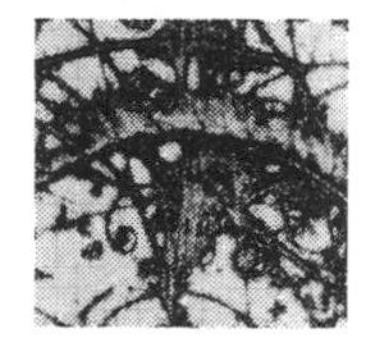

# The Particle Garden

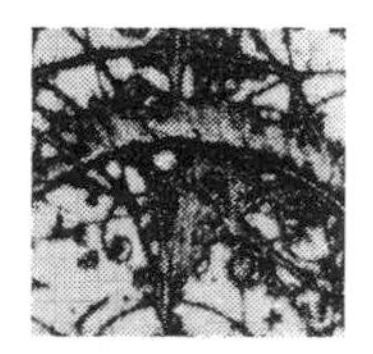

## Our Universe as Understood by Particle Physicists

*Gordon Kane*

*Helix Books*

BASIC BOOKS

A Member of the Perseus Books Group

Quotation by Lucretius on page xiii reproduced by permission of Penguin Books Ltd.
Art on page 34 reproduced by permission of Royal Astronomical Society Library.
Frank & Ernest cartoon on page 4 reproduced with permission from NEA, Inc.

**Library of Congress Cataloging-in-Publication Data**

Kane, G. L.

The particle garden: our universe as understood by particle physicists / Gordon Kane.

p. cm.

Includes index.

ISBN 0-201-40780-9 (hardcover) ISBN 0-201-40826-0 (paperback

1. Particles (Nuclear physics) 2. Standard model (Nuclear physics) 3. Nuclear structure. I. Title.

QC793.2.K365 1995

539.72—dc20 94-25804

CIP

Jacket design by Lynne Reed
Text design by Joyce Weston
Set in 10.5-point Usherwood by Shepard Poorman Communications Corporation

Second printing, April 1995
First paperback printing, May 1996

*To Lois*

# Contents

*Preface* *xiii*

*Prologue* *xvi*

Chapter 1 What Are We Made of? 1

*Atoms and protons have structure, but perhaps quarks and leptons do not. If not, they are the basic constituents of nature—and us.*

- Are Quarks and Electrons Nature's "Seeds"?
- "There Are More Things in Heaven and Earth . . ."
- What *Are* Quarks and Leptons?
- Unification
- The Role of the Theory

Chapter 2 A Brief History of Particle Physics 23

*We learned how to understand nature by starting with the simplest details, describing them mathematically, experimenting, and formulating a theory.*

- The Beginnings of Science
- The Beginnings of Modern Science: Measurement and Experimentation
- Three Paths to Today's Theory

# Preface

*One can laugh without being composed of laughing particles, can think and proffer learned arguments though sprung from seeds neither thoughtful nor eloquent. Why then cannot the things that we see gifted with sensation be compounded of seeds that are wholly senseless?*

*Lucretius,* On the Nature of the Universe
*(Translated by R.E. Latham, Penguin Books)*

Particle physics has made wonderful experimental and theoretical discoveries in the past three decades. While reports of these discoveries have aroused some public interest, explanations of them are difficult for two reasons. The obvious reason is the highly technical and mathematical nature of the work. The second reason is that research in progress is hard to describe. It is hard for experts who are used to thinking about the subject technically to write about it simply, and it is sometimes hard for less technical writers to write about it accurately.

As a professor at the University of Michigan I occasionally have had the opportunity to teach physics classes for nonscience majors. And during the years when the Superconducting Super-Collider (SSC) was widely discussed, I gave public talks about its goals and about science in general. From these activities, and from conversations with friends, I learned what probably should have been obvious—many people want to know the same sorts of things about particle physics that I want to know about history, art, linguistics, etc.: What is the field about? What are some of its great achievements? What are the exciting problems? We all want to know something about other fields, but not too much. So I have tried to achieve that balance.

Another kind of balance is needed between well-established and tested results, frontier topics under systematic study, and new speculative but unlikely thinking. Even the well-established particle physics is new, and that is what I have emphasized. I also include several major frontier topics—particularly Higgs physics and supersymmetry.

The first five chapters of this book are about what has already been learned, both historically and recently, and about the theory that incorporates the experimental and theoretical results. They don't require any previous particle physics knowledge. The sixth chapter reviews the experimental foundations of the now established theory, and the remaining chapters describe work in progress. Chapter 7 introduces some distinctions I have found to be useful when describing what "understanding" means to particle physicists. The chapters from 6 on do require particle physics knowledge, but all of it is provided by the preceding chapters. I hope that those readers who don't easily carry the information of the first five chapters along will still browse the rest, because it is possible to get from them an image of present and future directions of particle physics' explorations of nature, and a feeling for the kinds of questions particle physicists are asking today. A few figure captions present extra information.

Over the years I have developed some ways to explain ideas and results, and I have learned more ways from other people. I may need to apologize here to unknown persons: some of the examples and analogies and descriptions I use may be ones I learned from others, but long ago internalized. Certainly one person to whom I owe thanks for my now ingrained methods of both learning and teaching is my own thesis professor and friend, J.D. Jackson.

I came to write this book at the suggestion of Addison-Wesley editor Jack Repcheck. It is my honor to work with an editor whose intellect has caused him to originate the Helix series of books for the enrichment of science understanding. I thank him for his encouragement, without which this book would not exist

While writing this book I have benefited greatly from help provided by my wife, Lois. She is responsible for much that is good about the book, from the title to a great deal of any clarity and eloquence it may have. It was a pleasure to share this effort with her. I appreciate very helpful suggestions and comments from the many friends who have kindly taken time to read the manuscript, particularly Marty Einhorn, Nancy Elder, John Hilton, Chris Kolda, Eric Rabkin, Jean Rivkin, Jeff Rivkin, Bill Rosenberg, Marc Ross, and Jim Wells. I have also had helpful discussions with Sandy Faber, Len Sander, and Jack VanderVelde. Any errors or misstatements are, of course, my own.

# Prologue

1624 CE: Three scholars in Paris advertised that they would give lectures in support of the idea that matter is made of atoms. The authorities ordered that the audience be dispersed, confiscated the writings of the scholars, and prohibited any teaching about atoms under penalty of death. And seven years later all Jesuits were prohibited by their own leadership from holding or teaching that objects were made of atoms.

1906 CE: The Austrian physicist Ludwig Boltzmann was completely overcome by frustration and depression. For three decades he had tried, while often embroiled in unpleasant controversies, to convince influential European scientists that atoms exist. That year he committed suicide. Ironically only a few years later new evidence finally led to almost complete scientific acceptance of the existence of atoms.

1963 CE: Richard Feynman, the brilliant American physicist began his extraordinary three-volume *Lectures on Physics* with "If, in some cataclysm, all of scientific knowledge were to be destroyed, and only one sentence passed on to the next generation of creatures, what statement would contain the most information in the fewest words? I believe it is the *atomic hypothesis* (or the *atomic fact,* or whatever you wish to call it) that *all things are made of atoms. . . .*"

# 1
# What Are We Made of?

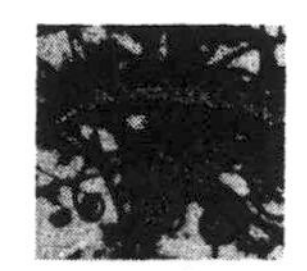

Particles, so tiny that one is less than insignificant, and in numbers large beyond counting, seed the garden that is our universe. A few ancient Greeks wondered what things were made of, and they guessed that the elaborate workings of nature depended on tiny elements. They coined the term "atoms" (*atomos*) meaning "uncuttable," and the name was given to the units now universally recognized by the symbol ·⚛ —a nucleus fenced by whirling electrons. Most people, if pressed, would probably concede that Adam was made of molecules that were made of atoms. That's true, but not the whole story. We now know that Adam's atoms were made of particles called quarks (sounds more like quartz than like quacks) and electrons, bound together by particles called gluons and photons.

To the satisfaction of physicists who had predicted its existence, in April, 1994 experimenters at the Fermi National Accelerator Laboratory reported evidence for the reality of the "top quark." In one sense their work brings to a successful close two and a half millennia of human efforts toward understanding how the natural universe works and what it is made of, because the top quark is probably the last particle of ordinary matter that will be found. To my mind the modern era of what came eventually to be called particle physics began in December, 1910, when

Ernest Rutherford carried out experiments which implied tha atoms had a structure consisting of a tiny nucleus surrounded b electrons. From that insight and the experiments it suggested one can directly follow a path that culminates in the Fermilal discovery of the top quark. The experiments Rutherford and hi junior colleagues carried out were the first ones in which ener getic subatomic projectiles were directed at a target, and th resulting subatomic particles detected and studied. Exactly th same approach was used to produce and detect the top quark but with a projectile energy about a million times greater thar that used by Rutherford.

It has been observed that there are two kinds of geniuses The first is smarter and somehow does things better than mos of us, but once we see their results we can understand how the got them; the second makes remarkable leaps at which ever their best peers can only marvel. Rutherford was the first kind Born in New Zealand in 1871, he excelled sufficiently to get scholarship to Cambridge University, and then professorships a McGill University, Manchester University (where he discovere the nucleus), and finally Cambridge University. He was loved an respected, and well rewarded. Sometimes great scientists ar singular people about whom fascinating anecdotes accumulate Not Rutherford—while there are some anecdotes, they tend t elicit at most a tentative smile. He was simply a fine, hardwork ing, effective scientist who made some of the most importan discoveries of the century.

In some ways the top quark discovery, and many of the con tributions (described in this book) to today's huge progress ir understanding how the universe works, are similar to Ruther ford's career. The top quark discovery is a remarkable achieve ment that required the integration of a variety of forefron techniques and approaches. It is not that top quarks have no been around; from the expected (and now probably measured properties of top quarks we can estimate that several times eacl minute a top quark is produced somewhere on earth, by collision of a cosmic ray from outer space with the nucleus of an atom in the earth's atmosphere. One top quark might be produced above Chicago in the atmosphere, the next above the Atlantic ocean. And top quarks are unstable, transforming into other particles in a tiny fraction of a second, so it is necessary to deduce their existence from finding unusual configurations of otherwise normal particles. The problem is to produce them in one place in a controlled way, and detect them. It could only be done if an accelerator of sufficiently high energy and intensity were available, a goal that was finally achieved after several upgrades at Fermilab. A detector capable of seeing the unusual configurations was finally constructed by a group of several hundred good experimenters, who had to work together successfully, perhaps not unlike the cooperation over time required for construction of Gothic cathedrals. Some people have lamented the fact that the frontiers of knowledge have receded so far over the past century that individuals can no longer make significant discoveries. I think it is more impressive that large numbers of good scientists can work together under difficult and challenging circumstances to make new discoveries about the basic properties of our universe.

Astonishingly, most physicists now believe that all the things we can see, from the smallest to the largest, are made of just three particles. One is the electron, the same electron that moves in wires when you use electricity. The other two are the "up quark" and the "down quark." Quarks are a lot like electrons. There are also particles in the universe that do not make anything we see—additional kinds of particles are needed to make a complete universe. I will describe the evidence for them in later chapters.

As the prologue suggests, in the past it was hard for people to accept that we and our world are made of tiny, passive "atoms." Today it is not so much religious or philosophical views that stand in the way, but the remoteness of what we are learning. New knowledge about particles has been accumulating since

*FRANK & ERNEST reproduced by permission of NEA, Inc.*

the mid-1960s and has been accepted by most physicists sinc the mid-1980s, but much of what is most interesting is not ver accessible. These tiniest units of stuff are studied by particl physicists, who normally describe them with mathematical nota tion rather than words. I am just one of these physicists to try m hand at explaining in words what we've learned about the sim plicity and harmony of the workings of the universe. I hop through this writing to help more people to see the patterns tha create nature, including all of *us*.

This is a conservative book. It describes work that is largel agreed upon and which will be the basis for future work, an then describes the goals of that future work. It emphasizes th view that science requires experiment to stimulate and verif any understanding of nature. There is an alternative approac which attempts to achieve dramatic purely theoretical break throughs and leap to a complete understanding of nature. happy rivalry exists between the "string theorists," who creat mathematical constructs and hope they match the data, an "phenomenologists," like me, who prefer to mix studying mod els based on data with suggesting goals to experimenters.

If the mathematical, top-down method works, our deepes understanding of nature could come swiftly, but the ploddin methods of phenomenology have gotten us to where we ar today. Perhaps it is like the tortoise and the hare. I would like th hare to succeed because haste may be necessary if the dee answers are to be known in my lifetime, but the tortoise has better track record so far. At any rate, both string theorists and phenomenologists rely on the particles and the theory described in this book for guidance in understanding the theories they are working with, and they hope for more guidance from experiments underway and yet to come. It is only in their method of attacking the unsolved problems beyond the horizon of this book that they differ.

I have tried to write this book so anyone with sufficient curiosity can read all of it. It doesn't require any mathematical sophistication beyond multiplication. It does get more abstract and speculative in the final chapters, where I describe current research on topics like "Higgs physics," "supersymmetry," and, briefly, the so-called "theories of everything" based on "superstrings."

The electron and up quark and down quark are actually those uncuttable "atoms" first imagined 2500 years ago by the Greeks—pointlike, indivisible particles from which the world around us blossoms. History gets in the way of appropriate terminology here; when, in the nineteenth century, the chemical elements (oxygen, carbon, silver, etc.) were found to each have a smallest recognizable unit, those units were called atoms. More research early in the twentieth century revealed that those atoms had structure (a nucleus at the center, surrounded by electrons). Although that meant they were not the indivisible atoms as conceived of by the Greeks, they got to keep the name. Then it was learned that the nucleus contained protons and neutrons and, in the 1960s, that protons and neutrons contained quarks.

Why do subatomic particles have peculiar names like "quark"? While searching for smaller and smaller constituents of matter, physicists have met things that have no counterparts in everyday life. We have to give them names in order to talk and write about them. We have leaned toward humanizing these unexpected discoveries with somewhat whimsical names, such as "flavor," "color," "charm," and "gluon." Often the names are suggestive of the properties that were discovered. For example, quarks have a new kind of charge, similar to electric charge Everyone knows there are two types of electric charge, calle positive and negative. The new charge has more types, and th rules for combining them are different from the rules for electri charge. For the new charge, three quarks, each with charge dif ferent from that of the others, are attracted to bind into a proto that comes out uncharged. Since that is reminiscent of the com bining of three primary colors to make white light, we call th charge "color." It has nothing to do with the colors of light, o course, but it is a useful way of communicating. Every one of th seemingly frivolous names physicists use has a precise mathe matical meaning. And also, the use of these words is a form o affectionate respect for the hidden workings of nature on scale we cannot normally access. Later we'll see how some of th names like "up quark," "down quark," and others came to b chosen.

## *Are Quarks and Electrons Nature's "Seeds"?*

Ordinary garden seeds contain genetic information that direct them to be broccoli or roses. The information that directs part cles is instead embodied in the interactions they can have wit other particles. Although each interaction is without purpose, th possible outcomes of interactions allow or disallow what ca occur in nature.

For example, consider speculative thinking. It requires speculative organ (such as the human brain) composed of bi lions of neurons each composed of thousands of molecules com posed of atoms. But the size of an atom is fixed by three basi quantities: the electron's mass, the strength of the interactio that can bind it to a nucleus, and the value of the quantity (calle Planck's constant) that sets the scale for quantum theory phe nomena. So the properties of the particles and the laws tha govern their behavior set a minimum size for atoms and thus fo a speculating organism. Because it is small, a butterfly can b beautiful but it cannot be given to speculation; a person can be both.

Organisms large enough to speculate are allowed by nature even though, because of their complexity, we could not predict their existence from a chain of deductions starting with quarks and electrons. Chance and time have conspired to create this world from what the constituents, interactions, and rules of nature permit. Wonderfully, it allowed thinking beings who not only speculate, but who speculate about their own origins and what they are made of, and who have conceived and value mighty notions like justice and morality, and who are beginning to think that it is they themselves who must tend their garden universe.

The method that has worked to learn what we and our world are made of, and how the parts fit together, is to probe more and more deeply into matter looking for structure, a process that began almost a century ago, with Rutherford's experiments. The goal is to find constituents that have no detectable structure—the fundamental constituents of matter. Consider a favorite painting, food, or person. Each is made of molecules. Molecules are combinations of atoms. Each atom has electrons bound to a nucleus. Each nucleus is made of protons and neutrons. And it turns out that each proton or neutron is made of up quarks and down quarks in appropriate combinations. Figure 1.1 illustrates the stages of matter.

Will it turn out that electrons and quarks are also composite? In recent years, two lines of reasoning, one experimental and one theoretical, have emerged that lead many physicists to believe this historical path has ended, and that electrons and quarks may be nature's seeds—the ultimate, indivisible constituents of matter.

## EXPERIMENTS FIND STRUCTURE, THEN NONE

Each smaller building block of matter that was discovered in the past was found by experiments that probed more deeply into

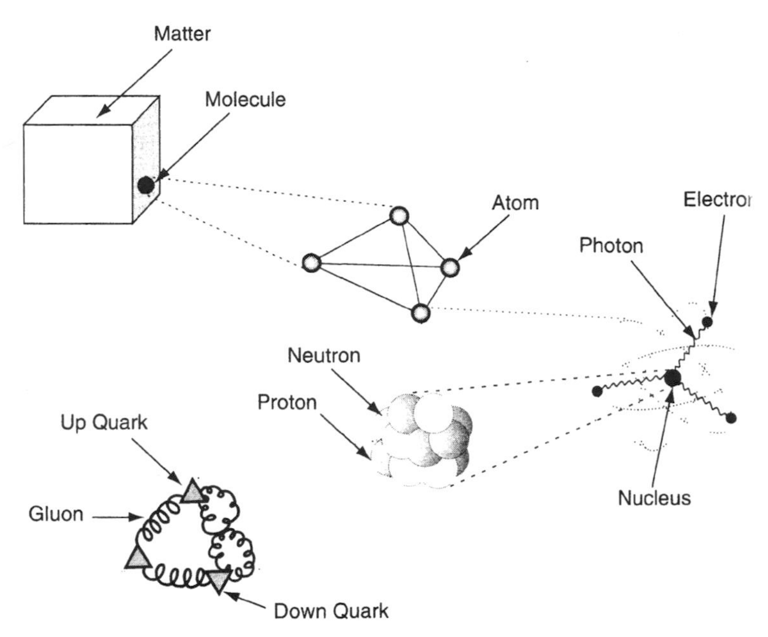


**FIGURE 1.1** *Matter is made of molecules. The molecules are combination of atoms (for example, salt is made of sodium chloride molecules, each mole cule containing a sodium atom and a chlorine atom bound together). Eac atom has a nucleus with electrons bound to the nucleus by photons. Eac nucleus is made of protons and neutrons bound together. Protons and neu trons are made of quarks bound by gluons. A typical small molecule has diameter of about one-millionth of a centimeter, a typical atom about a tent of that. A nucleus has a diameter only about 1/10,000 that of its atom. proton or a neutron is a few times smaller than a nucleus. Experiments hav demonstrated that quarks and leptons are at least 1000 times smaller tha protons, and they may even have no detectable size.*

matter, but recent experiments have found no deeper structure Probing more deeply into matter in physics usually means hittin an object with a more energetic projectile. For example, if yo throw a small steel pellet at a peach, it will bounce off. If yo shoot the pellet at high speed (therefore at higher energy) it will penetrate the peach, but bounce off the pit. So you can learn that peaches have pits without cutting into them. You could even learn the size of the pit by studying which pellets went all the way through and which ones bounced.

Historically, every significant increase in projectile energy has led to the discovery of a deeper level of structure. For over half a century larger and larger machines called accelerators (more about them in Chapter 5) have been built in order to probe matter with increasingly energetic projectiles—electrons and other particles. The most recent experiments on electrons and quarks have used projectile energies over a hundred thousand times greater than those at which structure might have appeared if history were going to repeat itself once more. Yet the electrons and quarks continue to look like impenetrable pointlike objects, with no apparent substructure.

There is another experimental way to learn whether things have structure. When atoms are hit by photons (the particles that make up light), occasionally a photon will transfer energy to one of the atoms. Electrons in that atom absorb the energy and are rearranged; we say the atom is left in a higher energy level, an "excited state." Typically the amount of energy needed for an atom to reach an excited state is a few electron volts. (That is a tiny amount of energy, about the amount that a flashlight battery gives to each electron it accelerates. When the flashlight is on, over a billion billion electrons are pushed through its bulb each second.) A nucleus is much smaller and more tightly bound than an atom, so it takes several million electron volts to excite a nucleus. A proton is even smaller and it takes a few hundred million electron volts to excite it. So if quarks and electrons had structure one might guess that they would be shifted into excited states with an energy transfer of several thousand million electron volts. But that has never happened in experiments, though attempts have been made with amounts of energy hundreds of

times greater than this rough guess. It appears that electrons an
quarks have no excited states. If that's true, it means that the
have no internal changeable structure.

The precise way in which an atom is "made up" of electron
moving around a nucleus, or a proton is "made of" quarks, i
complicated. Some "glue" is required to bind electrons to
nucleus, quarks to one another, and so forth. The "glue" als
turns out to be particles. The glue for binding electrons to
nucleus consists of the same photons that make up the light w
see. But the particles that bind the quarks have only recentl
been discovered; we call them "gluons." Gluons are particle
very much like photons, but gluons have some of the colo
charge mentioned earlier, and photons do not. Since the photo
and the gluons are particles too, we could say everything is mad
of electrons and quarks and photons and gluons, but it is usefu
to think of the electrons and the quarks as the matter particle
and the photons and the gluons as the binding particles.

### A THEORY IS REQUIRED

There is another and more subtle reason why the historic
search for smaller constituents may have ended. A complet
mathematical theory now exists that describes *how* things fror
particles to stars work *if* you start with electrons and quarks. It i
known as the "Standard Theory" of particle physics. The wor
"theory" usually has a precise meaning in physics, quite diffe
ent from its everyday usage. A theory is not a vague idea c
conjecture or argument, as in "His theory is that music . . ."
theory is a well-defined mathematical set of rules from whic
predictions can be calculated for a variety of phenomena. Th
goal of the process of explanation in physics is to find the theor
that incorporates and relates what is known.

Neither theory nor experiments alone can tell us how t
understand the real world. Experiments without a theory ar
hard to interpret and hard to relate to other experiments; theor
without experiment is just conjecture. The Standard Theory is

framework that unifies the new knowledge of recent decades with what has been learned about the physical world over past centuries, the achievements of Newton, Maxwell, Einstein, Heisenberg, and many more. The combination of experiment and the Standard Theory is what allows us to conclude that the quarks and electrons are the basic constituents of nature.

In this book I will first describe today's best theory of particles and their interactions, and then examine arguments that the theory can be made even better. Normally theories are formulated to explain the behavior of some set of objects, of some sizes, properties, etc. Theories become established as correct—that is, as describing the real world—by a combination of experimental tests and explanatory power. If the domain where the theory is formulated and tested is increased (toward smaller things like quarks or toward larger objects like the universe), the theory may then lead to incorrect predictions. Eventually an extended theory will describe the new domain, including more phenomena and explanations.

If we look toward the biggest things instead of the smallest ones, our conclusions about the basic constituents of matter are reinforced. We know that the earth and the moon are made of atoms. We have known for most of this century that our sun and even the distant stars are made of the same atoms and molecules that occur on earth. We learned that by comparing the colors of the light that reached us from stars with the colors of the light from atoms in laboratories. When an atom collides with another atom, as happens often in stars, it may absorb energy and get knocked from the its normal—or "ground"—state into an "excited" level. After a while it returns to the ground state, getting rid of the extra energy by emitting it as light (photons). The colors of the light emitted when this happens are different for each chemical element and characterize that element uniquely. From the moon to the largest things we can see, the universe is made of the same ninety-two natural elements we have on earth—each composed of electrons and a nucleus, each nucleus

made of protons and (usually) neutrons, each proton and neu
tron made of up and down quarks.

## *"There Are More Things in Heaven and Earth . . ."*

During the past half century we have discovered that there ar
even more particles beyond those that are the constituents o
what we see. Over past centuries we have learned that the eart
is not the center of the universe, but orbits the sun, and then tha
our sun is but a typical star like billions and billions of others
Now we have learned that many, perhaps most of the particle
in the universe, are ones that we are not made of.

### LEPTONS AND MORE QUARKS

The first to be found were neutrinos, ephemeral things that mov
forever, not bound by any force (so not bound into protons
nuclei, etc.), occasionally but rarely colliding with a quark or elec
tron. In addition there are four more quarks, called the charm
strange, top, and bottom quarks. There are also two more parti
cles, the muon and the tau, that are like the electron. Togethe
the electron, the neutrinos, and the particles like the electron ar
called leptons. They are created naturally in cosmic ray collision
and by experimenters at particle accelerators.

While the leptons and quarks may be indivisible, they ar
more interesting than the Greek "atoms" because under certai
conditions they can convert into other particles. The complet
set of particle interactions includes some that allow one particl
to make a transition into two or more other particles (with cal
culable probabilities). When a heavier particle converts into sev
eral lighter ones, the transition is called a "decay." The additiona
quarks and leptons do not occur in the things around us, becaus
they are unstable. They exist for only a very short time befor
they decay into up quarks and down quarks and electrons an
neutrinos. The muon is the longest lived of the additional quark

or leptons, and it lives only a millionth of a second before decaying into an electron and two neutrinos.

"Decay" is an example of an everyday word used in a special and precise way by physicists (without any of the usual negative connotations of the word "decay"). In physics, decay is just a way particles change into one another. Because of the interactions, particles can decay without in any sense having structure. The interactions themselves annihilate, or absorb, the initial particle and create the final ones. Physicists actually think of all interactions in this perhaps odd way. When a photon reflects off a mirror, physicists describe the reflection as an incoming photon being absorbed or destroyed by an atom in the mirror, followed by an outgoing photon being emitted (created) by an atom in the mirror. Even in collisions in which the type of particle changes, the same sort of description is used, with initial particles "destroyed" and final ones "created."

Surprisingly, though we do not understand the role these additional particles (charm, strange, top and bottom quarks, the muon and tau leptons, and the neutrinos) play in the basic scheme of things, we can predict with great accuracy every aspect of their behavior, from how likely it is that a given number of them will be created in any experiment or collision, to how long they exist before they decay into other particles, and what particles they decay into. From their properties we can see that these additional particles are intimately related to electrons and quarks, so that an eventual understanding of *why* there are electrons and quarks will have to involve the additional particles, even though a description of the things around us does not depend on them. From the viewpoint of the Standard Theory all the particles have an equal role—that we are made of some of them and not others is not important to the theory. In this chapter I am describing how the world looks from the point of view of human beings; in Chapter 4 I treat all the particles on an equal footing and ignore the bias introduced by keeping only our world in mind.

The additional particles may seem very different from a electron, which exists permanently if left on its own, while muon or a heavy quark decays in less than a millionth of a sec ond. Nevertheless, it is appropriate to categorize all of them a particles in the same sense, because each has a unique an unchanging set of properties that can be identified under all con ditions. Most of the properties have no counterpart in the every day world and they can be bewildering at first glance. A particl can be specified by its mass, spin, three charges (electric charge the "color" charge mentioned earlier, and another one we hav not yet discussed, the "weak" charge), and three number label (called "baryon number," "lepton number," and "family num ber"). These are all the charges or labels we know of toda Though it seems complex, once we have identified this set o properties for a particle, its behavior in all experiments can b described—we can predict how many will be produced in a ce tain type of collision, each particle's lifetime, and so on. (Thes properties and what they represent will be explained more i later chapters.)

## ANTIPARTICLES

There are additional particles in another sense too. In 1928 Pau Dirac proved that any theory that successfully integrated bot quantum theory and Einstein's special relativity (as the Standar Theory does) required every particle to have a partner. Particle are represented in the theory by solutions to the theory's equa tions. Dirac showed that if the equations had a solution for particle, then they also always had a solution for another particl with all charges opposite in sign, but otherwise identical. If a electron, with negative electric charge, a certain mass, etc existed, then another particle with the same mass, etc. but pos tive electric charge must exist. The result holds for all particles The partner was called an antiparticle. There are anti-up quarks antigluons, etc. For particles with no charge, such as the photon the antiparticle is the same as the particle so the distinction i ignored. I will always refer to specific antiparticles as "anti-" in this book, except for the antielectron, which was the first antiparticle discovered and got its own term, the "positron." A particle symbol with a bar on top represents an antiparticle; for example, e represents an electron and $\bar{e}$ a positron. There is a certain popular mystique about antimatter, but it is misplaced. Antiparticles are just particles, and which of a pair is the antiparticle is somewhat a matter of convention. I will just treat them alike.

An amazing thing is that the existence of antiparticles was predicted purely by a theoretical argument, and then they were found. There were really two remarkable aspects. One was that once positrons were recognized in experiments it was realized that they had actually been seen before but had been ignored because they were not expected to exist. Particles with opposite electric charges, such as electrons and positrons, curve opposite ways in a magnetic field. Particles curving the "wrong" way had been noticed in the 1920s, but because the observation did not make sense it was not taken seriously—recognizing that an anomalous effect is actually a new phenomenon takes great insight. In 1932 Carl Anderson at California Institute of Technology finally took them seriously and reported them. Seeing the report, other physicists quickly realized that they were the antielectrons (positrons) predicted by Dirac. The other remarkable feature of the antiparticle discovery was that human thinking guided by a theory could actually predict the existence of a previously unknown part of nature.

## DARK MATTER

There may be yet another kind of particles that do not occur in what we see around us. The additional quarks and leptons are not in us because any that are created quickly decay, in less than a millionth of a second. Antiparticles are not in us because any that are created quickly annihilate with their particle in a burst of energy. The third kind are called "dark matter." They interact

so little with other particles that they do not get bound int
protons or nuclei or atoms. We will see in Chapters 9–12 wh
they are thought to exist, and their connections to the othe
particles.

### OUR ELECTRONS AND QUARKS

The electrons and up and down quarks in us have mainly existe
since the first few minutes of the universe. The universe origi
nated in a hot big bang, where all particles (all the quarks, all th
leptons, photons, all antiparticles, dark matter particles, etc.
were created. As the universe cooled, all the heavier quarks an
leptons decayed into lighter ones, ending in electrons and up an
down quarks. The up and down quarks combined into proton
and neutrons. Some electrons and quarks are made when a pro
ton or nucleus from outside our solar system (an incoming pro
ton or nucleus is called a "cosmic ray") collides with a nucleu
in our atmosphere and creates particles, either electrons and u
and down quarks or particles that decay into them. Apart fron
these few, all the particles in us have been present in the univers
since the big bang.

## *What* Are *Quarks and Leptons?*

We don't know yet just what quarks and leptons are. They ar
not just tiny chunks, and there is no simple analogy to every
day things that can help us visualize them. We can characteriz
them but not explain them in terms of something else. Explain
ing something in physics normally means giving a quantitativ
interpretation in terms of other things that are themselves under
stood (analogies may help visualize but they are not explana
tions)—for example, explaining protons and their properties i
terms of their constituent quarks bound by gluons. Asking wha
the particles "are" is like asking what space and time "are.
We know their properties, we know their effects, but to sa

what they are will require new insights. We will return to this question in Chapters 7 and 11 briefly examining ongoing ideas about breaking the chain of explaining everything in terms of something else, and clarifying what "explain" and "predict" mean.

While there are still important questions to answer, it is astonishing how much has been learned in past decades. There is good reason to believe that after centuries of exploration, physicists have learned what we are made of. If you look inside worldly things you see atoms, and if you look inside atoms you see up quarks and down quarks and electrons, held together by gluons and photons. This statement is based on a century of experiments that take things apart, and a quantitative theory that describes the experiments and describes how to build things up again. Since things in the world are complicated, the theory to build them up necessarily leads to situations where the complexity prevents us from calculating the details. So long as experiments and observations that should be explained by the theory are indeed explained, and the range of such experiments and observations is judged to be sufficiently large, scientists will accept the validity of the theory. And an acceptable theory must not have any conceptual gaps—no need for a miracle to bridge a crevice. The so-called Standard Theory of particle physics satisfies those conditions.

## *Unification*

The world around us is a very complicated place, with many phenomena seemingly very different from each other. A rainbow, a blue sky, and a sunset may seem different, but actually they are all due to the scattering of sunlight, by raindrops or the atmosphere. The goal of physics is to explain and understand the natural world in terms of simple underlying principles. Taking things apart—the reductionist method—has led to great progress in describing nature. Physics has also been very successful at unifying—learning that apparently different phenomena reall were due to the same underlying mechanism.

The earliest unification was Isaac Newton's demonstratior over three hundred years ago, that motion in the "heavens" an motion on earth, claimed from antiquity to be unrelated, in fac obeyed the same laws. Many of the scientific achievements c the nineteenth century can be viewed as unifications. Once hea and sound seemed different, but now we understand both ar due to the motion of molecules. Once different forms c energy—energy of motion (kinetic energy), gravitational poter tial energy, heat generated by friction, chemical energy, mass and more—were each considered different things, but now w know they can all be converted into each other and the tota energy remains unchanged for any isolated system. The theorie of electricity and magnetism were unified by James Clerk Max well in the 1860s (about the same time that Charles Darwin pub lished the theory of evolution). Maxwell's unification generate a remarkable bonus when it turned out to explain light and optic too. It has often happened that a new theory led to entirely unar ticipated new understandings and predictions; presumably tha will continue to be true in the future.

Unifications occurred in chemistry and biology in the nine teenth century too. Before the theory of evolution the relation c different but similar existing species (such as humans and chim panzees) to one another was mysterious. Understanding tha they had evolved from a common origin led us to see anothe unity in the apparent complexity of the world around us. Evolu tion also unified our understanding of the world today with mi lions of years of prehistory. Although geological and fossil record of prehistoric times exist, and show a world and life forms differ ent from ours, evolution was the unifying concept that let u understand that what existed before is connected to us and to a contemporary species. Until the middle 1800s it was though that inorganic and organic chemical compounds, the latter th ones occurring in living systems, were separate and fundamen tally different; then it was learned that they were just different combinations of the same atoms, obeying the same laws.

In the twentieth century Einstein unified our concepts of space and time, and mass and energy. Quantum theory provided the rules to calculate the properties of atoms and molecules, which unified the domains of physics and chemistry. Astronomy and physics increasingly overlap. We have learned from the starlight reaching us that the earth and the sun and the stars throughout the universe all obey the same natural laws. Because light travels at a definite speed, light reaching us now from distant stars was emitted long ago, but can be recognized as light emitted by the same of kinds of atoms we have on earth, obeying the same rules of quantum theory that determine the pattern of colors of light from each atom. Starlight gives us experimental evidence that the laws of nature have not changed in billions of years.

## *The Role of the Theory*

What we are calling the Standard Theory is usually called the Standard Model, for historical reasons. In any area of physics there are initially a lot of models proposed by scientists trying to describe what is observed. Models are attempts to provide partial explanations for phenomena, and they guide research. A model in physics is a set of assumptions and rules, in mathematical language, that incorporates a certain amount of what is known and aims to relate what it describes to some new experimental or theoretical results. A model allows us to make predictions, with one of three results possible. A model can uniquely predict a value for something we know and get it right, which increases our confidence in the model and its other predictions. It may get it wrong, in which case the model must be modified or discarded. Or lack of knowledge of other quantities needed for the prediction may lead to a range of values including the known value. That result is encouraging and draws people into working

further on the model. As time goes on, people add to the mode
and it encompasses more explanations. At a certain, not pre
cisely defined, stage the model becomes accepted as a theor
(or else people lose interest in it, perhaps because some of it
implications are wrong). At some stage physicists start callin
the most successful model "the standard model" just to give it
name.

In particle physics through the 1970s the standard model, a
it was better tested and established, became accepted as the fu
description of nature on its level, and promoted to the Standar
Model. The Standard Model is no longer a model—it is now th
most complete and sophisticated mathematical theory there ha
ever been in the history of science. The only sense in which it i
a model is the historical one. In order to emphasize this distinc
tion I have called it the Standard Theory in this book, though i
most particle physics literature it is still called the Standar
Model.

While nature may have yielded some of its secrets about th
fundamental constituents of matter, it has also kindly and teas
ingly shown us, as we probe more deeply, not more structur
inside the electron and up and down quarks, but more quark
and leptons. Presumably the additional quarks and lepton
(muons, tops, neutrinos, etc.) have a deeper role to play in th
full theory, being somehow required in order to have a univers
like ours. Physicists hope the existence and properties of th
additional quarks and leptons will provide clues about why ther
are quarks and leptons at all. Neither the existence of additiona
quarks and leptons that are not in us, nor our lack of understand
ing of why they exist, prevents us from arguing that we under
stand what things in our world are made of. The theory canno
yet explain why the world is what it is, but it can successfull
describe the world as it is—including the behavior of all of th
quarks and leptons.

Though there are many ways to demonstrate things are false
a scientific approach to understanding the natural world ca

never prove an interpretation is true. Even though scientists know that in principle a new experiment could come along and show that previously accepted ideas are incorrect, in practice once there is a formulation with many experiments giving results that are interconnected by a theory, the implications of that whole formulation can be accepted. The experimental fact that no deeper structure is found beyond quarks and electrons held together by gluons and photons should convince us to take seriously the idea that we have been fortunate to live at a time when the final level of structure of matter was first seen. To really believe this, we need a meaningful theoretical framework. Some physicists hope that an extraordinarily complete theory will be formulated, one that incorporates all we have learned about the physical universe and tells us that many seemingly arbitrary pieces of the picture in fact must be the way they are. Such a theory is sometimes called a "theory of everything" (see Chapter 11, where a more modest name is used). By the end of the book what this means will be clearer, but for now let me describe it anecdotally.

In 1905 at the age of twenty-six Einstein published several extraordinary results that revolutionized physics. From then through the development of quantum theory in the 1920s he played a central role in the progress of physics. For the next thirty years, until he died in 1957, he increasingly focused his efforts on the attempt to construct a fully unified theory of all forces and matter. He had no success at all.

Today we know that Einstein could not have succeeded. He did not know about quarks, nor about the gluons that bind quarks. He did not know that the electromagnetic, weak, and strong forces seem to extrapolate so that they have the same strength at very high energies (Chapter 9), nor about ideas like supersymmetry (Chapter 10) that may be needed to make sense of unification. He could not know because experiments had not been done that would probe deeply enough to learn these things (Chapter 6). Today, while we do not yet have confidence that we

have the right unified theory (see Chapter 11) of all forces ir
hand, candidate theories exist that would have fully satisfied Ein
stein. They are being studied by experimenters and theorists. I
is the combination of the experiments that continue to show n
deeper structure, plus the apparent possibility that theorie
based on quarks and leptons and their interactions can be full
unified theories (even if we don't know which theory wi
emerge, the evidence is good that one will—see Chapter 9), tha
lead many of us to think the search for the constituents of matte
may have ended. It brings us a step closer to understanding th
universe, our place in it, and perhaps ourselves.

# 2
# A Brief History of Particle Physics

## *The Beginnings of Science*

About 2600 years ago a path of inquiry began that has led us toward a widely shared understanding of the natural universe, one that scientists from every country and culture are converging on. Along the way there were many deviations from the straight path, and at least once for a long time the whole effort nearly died out.

We know when and where this path of inquiry began, and why it happened there. About 600 BCE, in a few Greek city-states located on the western coast of today's Turkey, called Ionian Greece, a few people began to argue for the most important of all ideas about the natural universe—that the world around us was something we could understand, rather than a magical place where events could have supernatural causes.

A number of things combined there to create a milieu in which such an idea could originate and flourish long enough to take root. One was the invention of alphabetic writing. The first writing, invented before 2000 BCE independently in several places, was hieroglyphic, with each symbol representing a whole word or concept. Somewhat before 1000 BCE alphabetic writing was invented, on the eastern shores of the Mediterranean.

Alphabetic writing is more abstract than hieroglyphic writing
and pushes the mind of the user toward different sorts of logic
In Greece in 700–800 BCE vowels were added to the earlie
forms, achieving modern alphabetic writing.

Another important factor was the political situation in Ionia
Greece. The city-states were only loosely federated, each inde
pendent enough so that no effective means of imposing an
single social or political structure was possible. If someone'
ideas were not popular in one locale they could be presente
elsewhere. Without such freedom new ideas seldom survive. A
third ingredient was the location of the city-states at a crossroad
between a variety of cultures and languages at a time of increas
ing trade and prosperity. The residents of the city-states hear
about different gods. They met people who were used to doin
things in whatever way worked. It was easy to doubt traditiona
explanations and rules. The increasing prosperity of the citie
created a larger class of people with the leisure time to develo
and debate ideas.

The availability of the alphabet made writing and readin
much more accessible and, in particular, led to increased activit
in the writing of laws to regulate society. As people thought mor
about formulating a legal system, it was inevitable that analogie
with laws of nature would occur.

Organized religion was relatively weak in Greece. Rathe
they had a mythology featuring numerous gods who ofte
argued among themselves. A Greek who held alternative view
about the origins and workings of the world was not in muc
trouble.

Some of these factors occurred in other places, of course
and surely a variety of people in other cultures had simila
thoughts about the natural world. But in Greece the ideas too
hold, and a number of the essential aspects of what we call sci
ence were developed there over a period of several hundre
years. At least four were of great significance. First was the ide
that nature could be understood. Second was the conviction tha

mathematical language can and should be used to describe nature. Third, the notion of mathematical proof was formulated—it should be proved generally, for, say, all triangles, that certain properties hold, rather than stating what properties hold for some particular triangles. And, fourth, there arose the powerful insight that all matter is made of constituents—the Greek "atoms."

The Greeks naturally missed some things too, such as the idea of negative numbers. In fact they had no convenient notation for numbers. The clumsy and inhibiting Greek and Roman mathematical notation was replaced with today's system by the Arabs when they were the custodians of science in the Middle Ages. The concept of zero also came by that route after being first introduced by Hindu mathematicians. More important, though, the Greeks largely missed an essential set of ingredients of science—the notions of measurement and of testing ideas by performing experiments. They did not get past the stage of thinking that you could learn about how the world actually worked by simply making arguments about how it "should" be.

Later, as the Greek world crumbled, it no longer was able to support the doing of science. No other culture adopted science in the same way the Greeks had. There was essentially no scientific activity in the Western world from about 415 CE to the sixteenth century; effectively such work had come to an end there. When the Western world emerged from a millennium of intellectual darkness, scholars learned again about the ideas of the ancient Greeks, and took the first steps toward modern science. Some of the Greek ideas were transmitted via the Arab world, which had welcomed science more than any other culture in the intervening centuries.

Modern science might have developed even without the transmission of the Greek ideas because the circumstances in Renaissance Italy were similar to those in Ionian Greece. There were a number of nearby city-states, relative freedom of thought, people with time and resources, and so on. The development of

science was dramatic in Italy, until the Catholic church tried t
repress the work of Galileo. Italian science did not quickl
recover from that. But France, and England, and the Netherland
took over, and science took root in too many places in the sev
enteenth century to be at risk from cultural or political force
from then on.

Perhaps the crucial ingredient that guaranteed science woul
not die out after the Renaissance was the invention of the prin
ing press in 1454. In 1417 a copy of an ancient manuscript, "O
the Nature of Things," written by the poet Lucretius about 56 BCE
was discovered, the only copy ever found. It presents the idea
of Leucippus and Democritus, arguing that matter is made c
"atoms" in a way that seems remarkably sophisticated an
modern even today. It was one of the first books produced afte
the printing press was invented and it was widely distribute
throughout Europe, influencing many thinkers. When, in 154
Copernicus wrote his book arguing that the sun rather than th
earth was at the center of the solar system, efforts were made t
suppress his ideas, but too many copies had been printed an
circulated. It was not possible to control the dissemination c
ideas displacing the earth from the center of the universe. Th
printing press was the first technological development tha
played a major role in extending human freedom; its role i
modern times has been supplemented by the fax machine, sa
ellite radio, and TV that were major factors in extending huma
freedom in the 1980s and 1990s.

## *The Beginnings of Modern Science: Measurement and Experimentation*

The essential roles of measurement and experimentation bega
to be clearly understood only in the sixteenth century, with th
emergence of scientists such as Nicholas Copernicus, Tych
Brahe, Johannes Kepler, and Galileo. These scientific pioneer
tried to explain the regularities observed in the "heavens" an

to understand motion. Brahe's measurements of the motion of planets and stars, carried out near the end of the sixteenth century, had extraordinary consequences. He was the first person to believe in the importance of measuring natural phenomena as accurately as possible, because he was sure it would eventually lead to understanding them better—and it did.

A major impact of Brahe's measurements came when he saw a new star where there had not been one. It was a supernova, though of course he did not know that. He observed it carefully for a year and a half, from 1573 to 1574, as it became brighter and then faded away. Because he made precise measurements he could demonstrate that it was really a distant star in the "heavens" rather than a comet near the earth. Brahe's supernova was visible throughout Europe, so there was great interest when he reported that it was a distant star. Across Christian Europe everyone had been taught for centuries that the "heavens" were eternal and unchanging. Here was the first proof to the contrary. Brahe's claim was checked by a number of people knowledgeable about astronomy (just as all the results of science can be checked). This result had a major impact on freeing people's minds from the constraints of centuries of dogma about the natural world and how it worked.

Using Brahe's data, and partly freed from the prevailing world-view by Brahe's work, Kepler was able to show after long calculation that the orbits of the planets were ellipses rather than circles. The larger axis of a planetary ellipse is only a few percent longer than the shorter axis, so only with Brahe's new measurements could such a difference be detected and Kepler's results be established. Kepler was able to formulate the first truly mathematical law of nature, relating different observable properties of a planet's orbit. In a sense that moment, in 1609, was the beginning of modern science—once people learned that there was a formula that really described how a planet moved, there was no turning back from trying to describe everything in nature mathematically.

Some historians prefer to mark the beginning of moder science with the work of Galileo, which took place at about th same time as that of Brahe and Kepler. Galileo also worked ou formulas that described motion, but his were for motion on eart and came as a direct result of experimentation. Although it ma seem simple to us, the problem of how to understand an describe motion was one of the most difficult to solve—it too over 2000 years of work to get it right. Galileo carefully define velocity and acceleration, and he found the relation betwee them by repeatedly measuring the velocity and the acceleratio of a ball rolled down an inclined plane. That was the first tim a mathematical formula was deduced from planned exper mentation, rather than from observation of naturally occurrin phenomena.

Galileo may have been the first person in history to be full recognizable as a modern scientist (with the possible exceptio of some ancient Greeks). He understood that phenomena can b studied by analyzing them in terms of their simplest compo nents. He wanted as much as any modern physicist to make grand theory that encompassed all phenomena, but he unde stood as no one had before that a different approach, focusin on the simplest problems, proceeding step by step, and perforn ing experiments, was needed. With these ideas all the essenti aspects of a scientific method for studying nature were in plac From then on there has been a steady increase in the rate a which we have gained understanding of nature.

Since then, especially in this century, the temptation to fo mulate a grand theory of "everything" (Chapters 11 and 13) ha been overwhelming for some of the best scientists. Einstein, an later Heisenberg, spent years trying, in spite of clear evidenc that they did not have a firm foundation to start from. Once th Standard Theory was formulated, however, some theorists rea ized that perhaps the ingredients required to build a more unifie theory might finally be at hand. Indeed, now about 400 year after Galileo it has become possible to move from reductionisr (the process of figuring out, in increasing detail, what things are made of) to unification (looking at things as a whole). The search for a unified theory is now a legitimate research topic explicitly pursued by dozens of physicists, but if scientists had not followed a reductionist strategy for centuries we could not have gotten to this stage.

My purpose in this chapter so far has been to describe very briefly the way physics developed. The most important thing to understand is that people had to *invent* the methods of science. Science might never have happened. It is very fragile, requiring many factors such as the freedom to think about how the world really works, the resources to experiment, and the time to think and work.

## Three Paths to Today's Theory

### THE CONSTITUENTS

The process that led directly to the development of the Standard Theory and our present understanding of what we are made of thus began about 400 years ago. It followed three parallel paths: probing the constituents of matter, learning what forces determined how these constituents interacted, and finding the rules to calculate what happens. We can trace these paths up to the present, beginning with how we learned about the constituents.

In the early 1600s Pierre Gassendi, a French scientist, read Lucretius's manuscript and encouraged a number of scholars throughout Europe to do so too. Gassendi emphasized the need for experimentation to discover atoms, and wrote several books expounding his results and views. Robert Boyle, in England, did experiments studying how air in a tube was compressed under various conditions, and in 1662 he published results that could be understood if air were made of atoms. For over a century European scientists collected data about many different materials and chemical processes. Finally sufficient information accumulated that in 1808 John Dalton, in England, wrote a book showing that a very large number of results for chemical reactions could be understood if a number of chemical elements existed, each composed of its own kind of atoms. The search for elements proceeded through the 1800s, as increasingly many scientists accepted their existence. About 1870 the Russian Dimitri Mendeleev succeeded in organizing the known elements into a pattern, the Periodic Table. On the basis of patterns and regularities expected from the structure of his table, Mendeleev was able to predict the existence of several previously unknown elements and to argue that certain properties of others were mismeasured. When his predictions turned out to be correct, the idea that the atoms really existed gained considerable credibility.

About the same time as Mendeleev was working on chemistry, James Clerk Maxwell in England, and later the Austrian Ludwig Boltzmann, succeeded in explaining the properties of gases, including heat and temperature, based on the insight that the gasses were made of atoms. For example, higher temperature corresponds to faster motion of the atoms. This work convinced more people that atoms exist. Even so, because atoms were very small and could not then be directly observed, most scientists were not convinced.

Finally, in 1905, Einstein argued that a phenomenon discovered in 1827 by Robert Brown (and, therefore, called Brownian motion) actually provided direct evidence for atoms. If you look through a microscope at tiny grains of pollen in water, you will see that they move continuously and randomly. Einstein said the water molecules, being always in motion, were bombarding the pollen from all sides, causing its motion. And, most important, Einstein worked out theoretical formulas that involved the size of the water molecule, the pollen size, the temperature, and the distance the pollen would move in a given time. In 1908 Jean Baptiste Perrin, in Paris, showed experimentally that Einstein's predictions were completely correct. Finally, after 2300 years the existence of atoms was essentially universally accepted (only a few years too late for Boltzmann—see prologue). And for the first time the size of a molecule had been deduced—the diameter turned out to be about a hundred-millionth of a centimeter.

Scientists were not given much time to enjoy the great sense of contentment and beauty that can come from understanding part of nature. Already in 1895 the discovery of radioactive decay (and findings from other experiments) had begun to hint that the chemical atoms (the smallest piece of each element that was still recognizable as that element) were composites of other things, apparently containing in particular the particle we now know as the electron. In 1910, Ernst Rutherford and colleagues began a series of experiments aiming projectiles (from the radioactive decay of atoms) at gold atoms. It was those experiments that prompted the analogy that we could learn about a peach pit by shooting little projectiles at a peach and seeing them bounce back when they hit the pit at the center, but not when they went clear through. Rutherford's projectiles bounced back from atoms sometimes, and he was able to describe the effect mathematically and determine that atoms had a hard nucleus whose diameter was about 10,000 times smaller than that of the atom. The chemical atom was composite!

By 1926, with the development of the quantum theory that led to detailed explanations of the properties of atoms and the Periodic Table, it became clear that each chemical element had an atom whose nucleus contained a different number of protons, from one proton for hydrogen to ninety-two for uranium. Comparing that information with data on the weights of each element led to the insight that another particle must exist, one very much like the proton but with no electric charge. The neutron was finally directly detected in 1932.

All nuclei are made of protons and neutrons, and all chemical elements consist of atoms—electrons bound to a nucleus. Once the neutron was discovered, it was attractive to suppose that the basic constituents had been found and that all matter was made of protons, neutrons, and electrons. But at least three reasons soon emerged that suggested that view was unsatisfactory. First a prediction about the properties of protons and neutrons did no agree with experiments. Protons and neutrons behave like smal magnets. For any elementary particle the quantum theory predicts the strength of the magnet, but the proton and neutron behaved like stronger magnets than the quantum theory predicted, thus implying they were not elementary; eventually thei magnetic properties were explained in terms of quarks. Second after World War II the Stanford Linear Accelerator Center (SLAC was built, to probe protons and neutrons with electrons in experiments analogous to those in which Rutherford discovered the nucleus. The early experiments showed that the protons anc neutrons were composite rather than pointlike, though the projectiles (electrons) were not energetic enough to allow the experimenters to determine more about the actual structure. Third, a variety of experiments led to the discovery of a large number o other particles similar to the neutron and proton, collectively (including neutrons and protons) called "hadrons." When physicists imagine basic constituents, they have in mind that there will be only a few particles from which many others could be constructed. By the 1950s there were too many hadrons to believe some could be basic and the constituents of others. Simpler explanations for the large number of hadrons were sought.

Finally, in 1968, a new set of experiments at SLAC, utilizing a higher energy electron probe, discovered quarks in the proton and neutron. The signal was much the same as what Rutherford found. Just as in Rutherford's experiment, too many projectiles bounced back instead of going through, so they had to be hitting something hard like the peach pit. Richard Feynman and James Bjorken produced analyses showing how the recoiling electrons would behave if they were hitting pointlike particles (the quarks) and their predictions were experimentally confirmed. Also, in the mid-1960s Murray Gell-Mann and George Zweig had each argued that the existence of the large number of hadrons could be understood if they were made of quarks. Soon theorists deduced that the quarks imagined as constituents of hadrons and those detected at SLAC could be the same, and the quarks began to seem real.

In the early 1970s developments came rapidly. Theories based on quarks and leptons were shown to make sense in major new ways. The final breakthrough, the one that led to almost complete acceptance of the quarks, was the discovery of a new quark, the "charm quark" in November 1974, simultaneously at Brookhaven National Laboratory and at SLAC, by groups headed respectively by Samuel C.C. Ting and Burton Richter. Other members of the SLAC group, Martin Perl and collaborators, found the tau ($\tau$) lepton in the same data—another remarkable discovery.

### THE FORCES

I will turn here from the discoveries of particles and structure to trace the development of our knowledge of the forces. Before Newton the idea of force was not clearly formulated. People were of course aware that apples would fall to the earth if they dropped from a tree, and Galileo knew how to compute the path of a cannonball—though Aristotle did not (Fig. 2.1). People also knew about static electricity and about magnetism. Newton, however was the first to define clearly the idea of a force of attraction between any two objects, the gravitational force, and more generally make explicit the notion of force. Newton's beautiful demonstration that the motion of the moon (motion in the "heavens") and motion on earth (symbolized by the falling apple) were determined by the same laws had an impact on ideas far beyond its direct scientific consequences: it established that objects in the "heavens" were subject to the same laws as objects on earth, and suggested that the same laws of nature could describe everything, everywhere. If Newton's laws applied everywhere perhaps the universe was an orderly place rather than the haphazard one it seemed to be. Further, the apparent universality of Newton's laws suggested to some thinkers

**FIGURE 2.1** *The Aristotelian theory of motion predicted the path show for a cannonball. Galileo knew the path was a parabola.*

through the 1700s that other things such as political and huma rights should be universal.

Another important conceptual consequence of Newton' work came from attempts to understand how the force of gravit worked. While Newton explained fully how to calculate th effects of the force of gravity, the underlying mechanism wa mysterious, particularly because it seemed to require that th force act instantly over long distances; this was called the "actio at a distance" problem. Newton was aware that the lack of a understanding of the mechanism did not invalidate his formula tion of the laws of gravity and of motion. About 150 years passe before Michael Faraday and others began to understand in the 1820s that one should imagine that each body sets up fields throughout space—a gravitational field, an electric field if it had electrical charge, and a magnetic field if it was magnetized. Then every other body sensed the gravitational, electric, and magnetic fields and responded in ways determined by its own mass, electric charge, and magnetism. The action-at-a-distance problem was not fully solved until the work of Einstein early in this century, nearly two and a half centuries after it was recognized.

When Mozart wrote his music, around the time of the American Revolution, little more was known about electricity and magnetism than had been known to the Greeks. There had been hardly any experimentation on electrical and magnetic phenomena, and therefore little progress. Benjamin Franklin was one of the first people in history to experiment with electricity. A crucial breakthrough came in 1821 when the Danish scientist Hans Ørsted demonstrated that if an electric current flows in a wire a nearby magnetized needle will move. That result immediately proved that electricity and magnetism were somehow related. Over the next decade, André Ampère in France, Joseph Henry in America, and most of all Faraday in England explored the relations between electricity and magnetism and found a variety of phenomena giving a unified picture in terms of the electric and magnetic fields of each object and how these fields changed with time. Finally Maxwell, with extraordinary intellectual creativity, synthesized all that was known about electricity and magnetism into one set of equations (which we call, of course, Maxwell's equations), describing the unified electromagnetic force.

What does it mean to say electricity and magnetism are unified? What the experiments showed was that a moving electric charge gave rise to a magnetic field, one indistinguishable from the magnetic field of an appropriately shaped magnet, and that a changing magnetic field gave rise to an electric field indistinguishable from that of an appropriate set of electric charges.

Maxwell found a set of equations allowing all the differen electrical and magnetic effects to be calculated from one basi entity, the "electromagnetic field."

At this stage of history all the known forces (at that time, onl gravity and electromagnetism) had been given full mathematica form. Nevertheless, thoughtful physicists knew that the exis tence of atoms still had to be proved and that the forces in th atomic domain had to be studied, while at longer distances n one understood how the sun worked or precisely what the star were. It was also widely known that Joseph von Fraunhofer, i Germany, had looked at the light from the sun and stars wit prisms and seen special intense lines in the spectra. Others ha observed that when chemical elements were heated an observed with prisms they emitted light of particular colors. I atoms were structureless objects they should not emit ligh (structureless things can't do anything by themselves), particu larly a different and characteristic set of colors for each elemen If they had structure, what forces could be at work? Maxwe emphasized such problems in his *Encyclopedia Britannica* articl on atoms in the 1870s. In fact it turned out that the electromag netic force is what determines the behavior of atoms, but tha would not be understood until quantum theory was developed.

In the 1870s there was no hint of the weak or nuclear forc because those forces operate only at distances the size of th nucleus of the atom, and science had not yet probed that deepl The weak and nuclear forces also play essential roles in the work ings of stars, but in such a subtle way that it would not have bee possible to learn about them by studying stars.

Manifestations of the weak force were first detected in 189 by the French scientist Henri Becquerel. He noticed that a pho tographic plate stored near uranium became exposed. Initiall no one knew what to make of this phenomenon, but studies ove a decade by Becquerel and Marie and Pierre Curie and other showed that the effect did not diminish when the uranium ha been kept stored for a long time, and that a chunk of radium could heat a similar chunk of ice into steam in minutes while the radium seemed unchanged. Electromagnetism and gravity could not account for such effects, so a new force had to be at work. Properties of the weak force were only slowly learned. In the 1930s Hans Bethe showed it plays a crucial role in how the sun (and all stars) work. Many experiments from the 1930s to the 1970s gave results that led to a more comprehensive description of the weak force, but its theoretical status was unsatisfactory until it was incorporated into the Standard Theory in the early 1970s. In Chapter 4 I will explain more about the weak force, how it affects the natural world, and why it is called "weak."

Recognition of the nuclear force followed soon after Rutherford's discovery of the nucleus in 1911, and the understanding that each of the chemical elements had a nucleus containing a different number of protons. It became clear that some new "nuclear" force was working to hold the positively charged protons together against their electrical repulsion since like charges repel—gravity was far too weak to do that.

Thus by 1915 the weak and nuclear forces had been observed, so four different forces were known. After that, little changed about our understanding of forces until the Standard Theory was formulated in the 1970s, when three major developments occurred. First, the description of the weak interaction was formulated in such a way as to fully include the requirements of the rules of quantum theory and special relativity, rather than as the crude and inconsistent set of recipes it had been. Second, in much the same way that electricity and magnetism had turned out to be different manifestations of one underlying force (and had been unified into electromagnetism), the Americans Sheldon Glashow and Steven Weinberg, and the Pakistani Abdus Salam working in England, succeeded in showing that there was a sense in which the electromagnetic and weak forces were different manifestations of one underlying

force, the "electroweak" force. Finally, it became clear that th
force acting between protons and neutrons, the nuclear force
was really a derivative of the more basic strong force.

In everyday life most "forces" are effects, such as pushe
and pulls, of one part of the world around us on another part
After Newton, for physicists a force came to be defined as any
thing that could change the state of motion of an object. Th
concept of force has evolved even further since then. The mod
ern definition of force is any interaction between two objects. A
experimental observations can be accounted for by the fou
forces—gravity, weak, electromagnetic, and strong. The interac
tion can be as simple as a push or pull, but it can be much mor
complicated, perhaps involving the creation or destruction o
particles as well. The electromagnetic interaction can do severa
surprising things—it is attractive between objects of opposit
electric charge, but repulsive between objects of the same elec
tric charge. The electrical force speeds up a particle in the direc
tion it is moving, while the magnetic force pushes a particl
perpendicular to the direction it is moving. The strong force ha
almost no effect on quarks when they are very close togethe
but if they try to separate it captures them so tightly they essen
tially never get further apart than $10^{-12}$ cm. The weak interac
tion has no effects at all in everyday life, but at the subatomi
scale it turns electrons into neutrinos and up quarks into dow
quarks. All of these features are explained and predicted by th
equations of the Standard Theory.

## THE RULES

So far in this chapter, I have traced the development of ou
understanding of the constituents of matter, and of the force
that act on matter. Finally, we need to examine the developmen
of the rules for calculating the effect of the forces. The first state
ment of a rule was Newton's famous "second law." It says tha
any *object* (e.g., an atom, a particle, a baseball) of mass **m** wi
undergo an acceleration **a** if a *force* **F** is applied, and the *rule* t

relate these quantities is $\mathbf{F} = \mathbf{m} \times \mathbf{a}$. Without this equation, Newton's second law, we would not know how to calculate the acceleration even if we knew the mass of the object and what force was applied. Without knowing the force or the mass we also could not calculate the acceleration.

As time passed, other empirical laws were developed, but for over two hundred years there were no fundamental changes in our understanding of the rules. Then in the early twentieth century Einstein pointed out that for objects moving with speeds close to the speed of light, Newton's law would not give the correct results; the requirements of Einstein's "special relativity" must be added. Around the same time the problems with understanding atoms showed that Newton's law and Maxwell's equations both failed at very small distances. In the 1920s new rules, embodied in a structure called the "quantum theory," solved the problem of understanding atoms. According to those rules, a new equation, the Schrödinger equation, replaced Newton's law. Finally in the late 1920s Paul Dirac wrote the "Dirac equation," combining the constraints of special relativity and of the quantum theory.

Today there are modern versions of the rules that are technically much more powerful and enable us to solve the equations much more effectively, but the foundations were in place by the 1930s. One can understand the evolution of the rules by imagining various limits. Starting from the Dirac equation, and going to the "nonrelativistic limit" where no particle in the system is moving with a speed near the speed of light, results in the Schrödinger equation of quantum theory. From that stage one can go to a situation where quantum effects are not important by considering only problems where nothing is as small as an atom, and then Newton's law results. Newton's law is not incorrect, but it only applies in a limited domain of velocities and sizes. Its domain is large, running from atomic sizes to galactic sizes, and covering all speeds achieved by objects on earth before this century. It has happened often in the growth of science that earlier results are incorporated and extended into a more encompassing result. What is new about the Standard Theory is that it applies at all possible speeds, and out to the edge of our universe, and back to within a tiny fraction of a second after the beginning of the universe. In that huge domain it is here to stay. Nevertheless, we still expect the Standard Theory to be extended and incorporated into a still more basic theory, as we will see in Chapters 7–11.

Quantum theory and special relativity were astonishing discoveries, and led to rules very different from those they replaced while still encompassing the earlier ones. Consequently historians and philosophers characterized them as revolutions, and described the history of scientific development as periods of normal science followed by revolutions. Actually, the discoveries of the particles and forces over the past century do not fit well into this view of history. Instead, there has been a steady process of development of experiment and theory, with each stage suggesting the next ones. There was no discontinuous or revolutionary stage, except possibly the very notion that we may finally know the constituents and their interactions.

# 3
# Doing Particle Physics

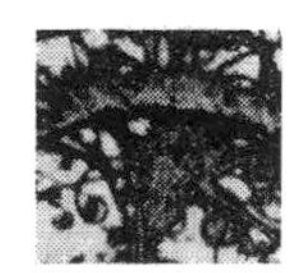

Never Frankenstein (fiction is truly stranger than fact), rarely Einstein, particle physicists seem finally to have shaken off the old absentminded professor stereotypes, and now instead seem to be benevolently, if quizzically, regarded as contributors to contemporary culture. Older theories like special relativity and newer ones with ideas like dark matter give the thrill of the unknown and the possible. Day-to-day life, though, includes the rigors of education and the worse rigors of getting funding. The Superconducting SuperCollider (SSC) project, with the biggest funding of all, was conceived in 1982 but dismally terminated in 1993. It brought particle physics wide attention and excited public imagination; its curtailment remains as a symbol of sadness and apprehension for those who would adventure beyond the edge of the known.

## *Education*

Students of particle physics are explorers. But instead of being to the north, south, east, or west, their frontiers are to the small—the particles that form our world—and, more recently, to the large—unseen particles that possibly fill all space as dark matter. Because they are searching for the unknown, we aim to teach them a combination of flexibility and rigor—an ability to see and comprehend what has not been seen or analyzed before.

Becoming a particle physicist is a process that takes severa years of training and apprenticeship after the bachelor's degree Some physics students enter graduate school unsure of which specialty they will choose (atomic physics, materials physics etc.) and gravitate to particle physics, but most aimed them selves in that direction from early on. Somewhere around the time I began high school, for example, I came across the fine little book, *The Universe and Dr. Einstein,* by the writer Lincoln Barnett. (Its front cover says it's "A clear, concise, and fascinating explanation of Einstein's theories, how they were developed, and how they have changed the course of modern thought.") From that time on I knew that I wanted to study those ideas. But when I encountered undergraduate physics courses in college they seemed to have little to do with understanding the universe bet ter. There were levers and capacitors and lenses, while I wanted to know what we are made of and what our place in the universe really is. Eventually I learned from wise professors such as Victo Weisskopf and Francis Low, both at the Massachusetts Institute of Technology, that there was a connection—physics was a highly unified structure and you needed to come at it from all sides to make sense of it.

Students begin to participate at the fringes of particle physics research early in graduate school. By the third year they are ready to work almost full-time on research. Fewer and fewe people can know enough both to do experiments and to grasp the increasingly complex theory. Probably the last person exper in both was Enrico Fermi, working mostly in the 1950s, afte whom Fermilab (and considerable physics) is named. The learn ing path is rather different for experimenters and theorists.

Experiments necessarily need equipment at the forefront o technology, and this equipment is usually large and complex and time-consuming to build. Today most particle physicist explore the unknown with huge "microscopes" that are the accelerators and their detectors. Some, though, fly balloons carrying detectors high into the atmosphere, or place them deep underground in tunnels and mines, to look for dark matter or proton decay or solar neutrinos.

Ideally, experimental physics students help with detector design and construction, along with a number of other graduate students, "postdocs" (recent PhD's), technicians, engineers, and senior physicists, from a number of universities and labs. Most experiments take place at accelerators located at a few national laboratories. When it is ready, an accelerator detector is placed in a beam of particles, and the resulting collisions are studied. A "run" usually lasts for several years, perhaps with upgrades to improve the ability to study some interesting, unexpected phenomena. To keep everything working right and to quickly repair problems, a number of people have to be available, so experimenters "run shift" twenty-four hours a day, seven days a week, as long as the beam is on.

After data is taken, it has to be studied and analyzed to see how much can be understood in terms of existing knowledge, and what—if anything—is new. To actually find something new requires wisdom and luck as well. Extensive computer software is needed to relate what the detector records to the physics interpretation, and the student is involved in that analysis too.

Eventually the graduate student gets a PhD, usually six or seven years after beginning graduate studies. Some leave basic research to take jobs using the high technology and analysis skills they have learned; they go to high-tech companies, to computer hardware and software companies, to government labs doing other kinds of physics, and even to Wall Street firms and similar places where analysis and applied mathematical skills are valued. Those who stay in particle physics normally become postdoctoral associates. The postdocs have, scientifically, a wonderful situation. They can concentrate full-time on physics goals, having mastered the essential techniques as students, undistracted by the necessity to raise and manage funds to pay for the

experiments as senior experimenters must. There is a downsid
though. Because the actual data is taken at central laboratorie
while jobs are mostly at universities, the experimenter mus
spend considerable time away from his or her home and famil
Eventually, probably now in their early thirties, those wh
stay in the field hope to get positions as university faculty o
senior scientists at laboratories, with new opportunities an
responsibilities.

A theory graduate student, on the other hand, will wor
under one faculty member, first doing one or two relatively sin
ple problems to test his or her abilities, and then focusing on
single detailed problem at the frontier on which he or she wi
concentrate full-time for about two years. Typically the theor
student earns a PhD in four or five years, followed if possible b
a postdoc job. Young theorists do not normally work "for" an
one while they are postdocs. They are on their own, though the
often collaborate with other theorists. It is during that postdo
period (two to six years in most cases) that the young theorist i
expected to mature into a particle theorist who independentl
can choose research directions.

In a better world it would be possible for anyone with th
talent and motivation to study whatever field excites him or he
and then get a job teaching and working in that area. Two cer
turies ago essentially no one could; today some but not all car
Getting the first postdoc job for a particle theorist is, in practic
largely up to the professor who was the doctoral candidate's Ph
thesis advisor. Those doing the hiring at other universities an
laboratories know the advisor and his or her work, while th
student is unknown. I am motivated to get good positions for m
students, but I am at least as motivated to maintain a good rel
tionship for many years with colleagues elsewhere. Unlike man
other academic and nonacademic areas, most particle physicist
in the world know one another, both because the field is a sma
one, and because communication and travel are an importar
part of the research process.

When the time comes to apply for a second postdoc, or a faculty job, the young theorist has had the opportunity to demonstrate his or her abilities; hiring decisions can be based more directly on the postdoc's work. If there are enough jobs to go around, the decisions are fairly objective. In the 1990s, because of continual cutbacks in the funding of particle physics (only a few percent a year, but they add up), the number of openings for postdocs is less than it has been for some time. The number of faculty openings is also very small, only a few each year in the United States. Theorists who seem to be as good as those already holding positions at top universities are being forced to leave the field. Under those conditions postdocs need mentors in order to compete for positions.

In the United States today there are, as someone has joked, about twenty-five universities and laboratories that make up the Top Ten. A student or postdoc can work at any of them and have access to the resources and colleagues needed to compete at the forefront of particle physics. At another ten to fifteen places it is possible to compete, but harder. There are somewhat over a hundred positions for postdocs in particle theory, with perhaps half available each year. There is some trade back and forth with Europe, and a little with Japan. In recent years physicists from the former Soviet Union and Eastern Europe have competed for the U.S. positions in unprecedented numbers too. There are about two hundred well-qualified theorists applying for the fifty or so postdoc positions each year, and almost that many for a small number of faculty jobs each year. Since many applicants are well qualified, other factors (including mentors, personal interactions, how fashionable the applicant's work is) play a significant factor in determining who actually gets the jobs.

Some difficult questions are raised by these numbers. It is easy to say that students should not be trained in a field with a job shortage. But when faced with a student who wants more than anything to understand better how the universe works, and to participate in the quest for more knowledge, it seems the wise thing to do is to explain the risk but then to let the student finally decide. Some are dissuaded, most not. Most professors accomplish more when they have students to assist them in their research, so it is not clear professors are objective about these issues.

The next challenge in the United States is tenure. Young faculty members will be evaluated, normally in their sixth faculty year, for a tenure position. Their research productivity and leadership potential are what is mainly evaluated, but their teaching effectiveness is included in the criteria. This time can be extraordinarily challenging for young people—they are often starting a family, buying their first house, learning to teach, and finally having an opportunity to achieve their own research goals. Over two-thirds eventually get tenure at some university. Once they become professors at universities they normally spend about half of their total time teaching—more during the academic year, less in the summer—and half doing research. They teach both undergraduate and graduate students.

## *International Community*

A very effective culture for doing particle physics has developed over the past half century, a culture that stresses the value of communication and the international nature of science.

Particle physics is done in many countries, particularly in Europe, the United States, and Japan. Counting only the very active particle physicists, there are several hundred theorists in the world, and perhaps twice as many experimenters. Including less active physicists, there are a few times more. After a while most active researchers know each other personally. Every major university and laboratory has (typically) a weekly seminar in experiment and one in theory, usually given by someone passing through or brought in for that purpose. The visitor describes his or her work, which is often not quite finished. Discussion and (often heated) argument follows. The visitor in turn hears about what local people are doing. In recent years electronic mail has provided a new mechanism that facilitates and stimulates communication. Altogether there is a valued camaraderie generated by sharing the same history and acquaintances and goals (in spite of sometimes fierce competition).

## *Funding*

Since the results of forefront research are not likely to have applications soon, essentially only governments can fund big science. Traditionally the United States has funded research in particle physics and related sciences in order to be preeminent in those fields, and to assure economic and military strength, even though particle physics results have no direct military applications.

More specifically, during the decades since World War II, it has come to be understood by economists and policy leaders that there are several reasons, in addition to national prestige, for a country to fund basic scientific research, even research that is unlikely to have immediate applications. Not only does the new knowledge add to our present cultural heritage, when future historians look back at the twentieth century advances in particle physics will stand out as one of the great contributions to the intellectual history of all time. This intellectual excitement, if it continues, attracts new young people to science and engineering careers; that keeps the cycle going as many of the young people go to industry with a state-of-the-art understanding of new techniques and ideas (the best method ever invented for "technology transfer").

Strong science exists only in economically strong countries for two reasons: first, because science requires funding, and second, because the findings of science stimulate the countries' economies and make them economically stronger. The results of research often lead to applications that improve quality of life and economic options. Although stimulating the economy is generally not the immediate goal of scientific projects, the nev techniques and devices required by forefront research often pro duce major high-tech and economic impacts, or "spin-offs." Often the equipment required to do large research projects pro vides a short-term guaranteed market that can turn start-up com panies into the viable industries that also stimulate a health economy. That role was played by high-tech military procure ment in the United States from the 1950s through the 1980s.

The main responsibility for funding the large particle physic laboratories in the United States (Brookhaven, SLAC, and Fermi lab—see Chapter 5) has been given to what is now the U.S Department of Energy (DoE). DoE also funds most particle phys ics research at universities. The National Science Foundatio (NSF) funds the Cornell University electron collider, and a signif icant part of all university-based research.

The whole funding system is based on peer-reviewed pro posals. Every year the laboratories themselves, groups of physi cists at universities, and individuals submit proposals to DoE an NSF for research funding. All are reviewed by experts, and finall the DoE and NSF scientists (all have physics PhD's) arrive at budget, based on the advice they receive and on the total amoun they are allowed to allocate.

Theorists need funding to support students, postdocs, com puting, travel to meetings, secretarial help, and the visitors an communication that are virtually their laboratories. For experi menters there is another aspect. To carry out an experimen using accelerated beams at a laboratory, experimenters mus prepare a proposal to the laboratory, giving scientific and techni cal justification. Each laboratory has its own scientific "progran committee," composed about two-thirds of experimenters fron other institutions, usually including some from Europe an Japan, and one-third theorists. Every proposal is considere using criteria such as its scientific importance and its technica feasibility. Manpower and funding criteria may also be included but they are secondary. The program committee is always advi sory to the laboratory director, who normally approves an experiment that is endorsed by the committee, but may occasionally not do so, and who could approve one not endorsed by the committee if he or she thought it justified.

Once an experiment is approved, it has a go-ahead for detailed design and construction of a detector, assuming its funding proposals were also approved. Before the experiment is scheduled to take data the director must be convinced it is ready, and the program committee may be consulted again.

As detectors have grown in size to the huge ones needed at current and future colliders, there has been some modification of this procedure. The detectors are so large and technically challenging and expensive that a large number of experimenters (a few hundred for existing ones, more in the future) must join together to build them. Only one or a few such large detectors can be used at each collider. Initially several may be proposed, but after the review process some are not approved. Usually on approval the laboratory makes a commitment to provide some funding from its own support. Upgrades, modifications, and future running times may still be reviewed by program committees.

The U.S. funding and peer evaluation system has many strengths, and is rather effective and efficient at maximizing good physics, but it has two major weaknesses. First, all funding commitments are for one year only, but the real time scale for doing particle physics projects is three to four years or more. Second, the overall scale of the budget is not fixed by scientific criteria or by taking optimum advantage of opportunities.

Particle physics developed differently in Europe. After World War II the European countries were able to agree on a plan for a European laboratory for particle physics research, called CERN after its French acronym, Centre Européen Pour la Recherche Nucléaire. (CERN has two official languages, French and English; in practice, all physics seminars are in English because it is the only language common to physicists from many countries.)

CERN has stable funding at a fixed level, always corrected fo
inflation. Because they know ahead of time what their fundin
level will be, they can plan future construction efficiently. Eac
member country of CERN contributes in proportion to its gros
national product (with an upper limit). For some years there wer
thirteen member countries. Recently more have joined, makin
nineteen at present. The total CERN budget is about the same a
the total U.S. budget for particle physics (there is some ambiguit
due to exchange rates).

In addition, each European country supports particle physic
at its universities and institutes, and some have experimenta
laboratories. Germany has a large laboratory, France and Ital
smaller ones. England had its own laboratory through the 1970
but has discontinued its use for particle physics except as
staging place for English experimenters working elsewher
Combining the CERN budget and the funding in the Europea
countries, the support for all of particle physics in Europe i
almost twice that in the United States, even though the Europea
population and GNP is only about 25 percent larger. If the Unite
States had built and operated the SSC, the total U.S. funding fo
particle physics would have been nearly as large as the Europea
funding.

## *Au Revoir, SSC*

By the end of the 1970s the Standard Theory was becomin
well established. So in 1982, particle physicists began a series o
international studies to discuss what facilities were needed t
test the Standard Theory and ideas beyond it. Almost yearly sev
eral hundred physicists met for about three weeks to discuss an
evaluate possible directions. It soon became clear that it was pos
sible to plan a facility that could guarantee progress, and that suc
a facility was technically feasible. Less powerful facilities coul
be proposed, and might lead to important results, but also migh
not. That ideal facility came to be called the Superconductin

SuperCollider (SSC). The particle physics community took the attitude that it would be asking for a lot if it asked for the SSC, but it decided to ask and let the government answer.

In 1985 the particle physics community submitted an initial proposal for the SSC to the DoE. After reviewing it, the DoE in turn went to the Reagan administration, which finally said to go ahead and make a preliminary design and cost estimate; some funds were allocated to do that. According to that design and estimate, necessarily tentative because no funds were available for prototypes, and because no specific site had been chosen, the SSC would cost a little over three times what Fermilab had cost, calculated on the same basis and in dollars corrected for inflation. That amount was about $3.3 billion for the bare accelerator, or about $4.5 billion when the cost of detectors, computers, user facilities, set-up operations, and so forth, were included, all in 1986 dollars. Approval was given, a site chosen.

The next stage was a detailed site-specific design. After careful study, one change was made in the design of the magnets that increased the cost of the machine itself about 15 percent. The final cost estimate for the machine plus all associated costs was completed in 1990. It was reported in then-year dollars (each year's expenditures in that year's inflated dollars) rather than 1986 dollars, and came to $8.25 billion, which was the same as the 1986 cost except for the one design change and the effects of inflation. No further design changes or cost increases of any size occurred, though after the Bush and Clinton administrations extended the construction period the cost in then-year dollars of course grew with inflation.

During congressional debates SSC opponents and journalists made repeated claims of cost overruns. But in sworn testimony on the SSC before the Senate, in the summer of 1993 shortly before the final vote that led to its cancellation, the Secretary of Energy said that the SSC was "on budget and on schedule." And a statement of the executive board of the American Physical Society, representing all fields of physics, said in September

1993, "The Supercollider is a project of great scientific merit that has met each of its technical milestones."

With the SSC, particle physics would have been in an exciting position, beginning by 1999 to gather the evidence that had been hinted at in developments of the 1980s, if Congress had appropriated the funding promised when the SSC was approved in 1989. Without the SSC particle physics is not dead or even in critical condition, though several years will be needed for many physicists whose careers were badly disrupted to recover. Instead, particle physics is in the same position as most fields of science in the United States today, operating somewhat below a healthy level, unable to take advantage of most of its exciting opportunities, doing the best it can. We will meet many of the exciting opportunities in the following chapters.

# 4
# The Standard Theory

So far I have presented the surprisingly simple picture particle physicists have been able to draw of the fundamental seeds from which all of nature grows, followed by a brief view of the historical routes leading to the Standard Theory. In this chapter I will change perspective and summarize the Standard Theory without regard to its history, or to the special status some particles and forces have for us and our everyday world. From a fundamental point of view the fact that there are lots of electrons and no tau leptons in people is irrelevant—the two particles have an equal status in the theory. Similarly, the weak force plays no role in the human body but to the fundamental physics it is as important as the electromagnetic force. Quarks and leptons were around from the beginning of the universe, interacting via all the forces. The garden grew people only relatively recently, billions of years after the seeds appeared.

For physicists the word "theory" implies a basis in experimentation. A theory embodies many crucial experiments in its formulation, and so the Standard Theory is not just a theory but the combination of mathematical expressions and their experimental foundation. It builds in all the natural laws we know (except gravity). In physics, once the decades of development of a new stage of understanding of nature are completed, it is

customary and appropriate to formulate the theory in a concis
mathematical way, ignoring the often tortuous routes traverse
on the way to the final result. The Standard Theory, the uniqu
result of many ideas and experiments arrived at over severa
decades, contains a number of unanticipated elements.

The Standard Theory is described in terms of the particles o
matter (quarks and leptons) on which forces act, plus the parti
cles that transmit the forces (of which there are two kinds, gaug
bosons and Higgs bosons). The gauge bosons include the photon
which transmits the electromagnetic interaction, and simila
particles for the other interactions (gluons, W's, and Z's); thes
particles will be described shortly. In particle physics the word
"force" and "interaction" mean essentially the same thing
According to the Standard Theory the interaction of Higgs boson
with other particles is responsible for the existence of the mas
of each particle. We do not yet know whether Higgs bosons actu
ally exist, but we do know that if they do not exist then som
other (presently undiscovered) phenomenon must exist tha
plays the same role as the Higgs bosons. The rest of this chapte
elaborates on all of the particles and interactions introduced i
this paragraph, except for the Higgs bosons, which are describe
in Chapter 8. Table 4.1 (which, for easy reference, is reproduce
inside the back cover of this book) summarizes the Standar
Theory from one perspective.

## *Matter Particles*

Matter particles (quarks and leptons) occur in certain patterns
Figure 4.1 will help to visualize them. There are six flavors (yes
this choice of names is based on the analogy with ice cream) o
leptons: the electron (e), the electron neutrino ($\nu_e$), the muon
($\mu$), the muon neutrino ($\nu_\mu$), the tau ($\tau$), and the tau neutrin
($\nu_\tau$). The six flavors can be separated into three families that ar
almost copies of each other: the electron family, the muon fam
ily, and the tau family. Two different features tell us that th

**TABLE 4.1**

| | *NAME* | *MASSES*** | *FEELS FORCES** | *MEDIATES FORCES** | *ELECTRIC CHARGE* | *COLOR CHARGE* | *SPIN* | | |
|---|---|---|---|---|---|---|---|---|---|
| $e, \mu, \tau$ | charged leptons (electron, muon, tau) | 1/1836; 1/9; 1.9 | EM W | — | -1 | no | 1/2 | leptons | fermions |
| $\nu_e, \nu_\mu, \nu_\tau$ | neutrinos | $<10^{-8}$; $<1/3500$; $<1/30$ | W | — | 0 | no | 1/2 | leptons | fermions |
| u, c, t | up, charm, top quarks | 1/235; 1.6; 165 | EM W, S | — | +2/3 | yes | 1/2 | quarks | fermions |
| d, s, b | down, strange, bottom quarks | 1/135; 1/6; 5.2 | EM, W, S | — | -1/3 | yes | 1/2 | quarks | fermions |
| $\gamma$ | photon | 0 | none | EM (binds electrons and nuclei into atoms) | 0 | no | 1 | gauge bosons | bosons |
| $W^{\pm}$ | weak bosons | 85 | W, EM | W | ±1 | no | 1 | gauge bosons | bosons |
| Z | weak boson | 97 | W | W | 0 | no | 1 | gauge bosons | bosons |
| g | gluons | 0 | S | S (binds quarks into hadrons†) | 0 | yes | 1 | gauge bosons | bosons |
| h | Higgs boson | not known; 65-160$^{\Delta}$ | W | generates mass | 0 | no | 0 | | bosons |

* EM = electromagnetic, W = weak, S = strong All particles feel the gravitational force, but it is neglected for the Standard Theory since it is much weaker than the other forces. Photons do not feel the forces directly, but they interact with all electrically charged particles

† Proton, neutron, lambda, pions, kaons, and others are collectively called hadrons: proton, neutron, and others made of three quarks are called baryons, while pions, kaons, and others made of a quark and an antai-quark are called mesons

** Ratio to proton mass; < means "less than"; quark masses only accurate to about 20%

Δ If a fundamental Higgs boson exists it is expected to be in this range

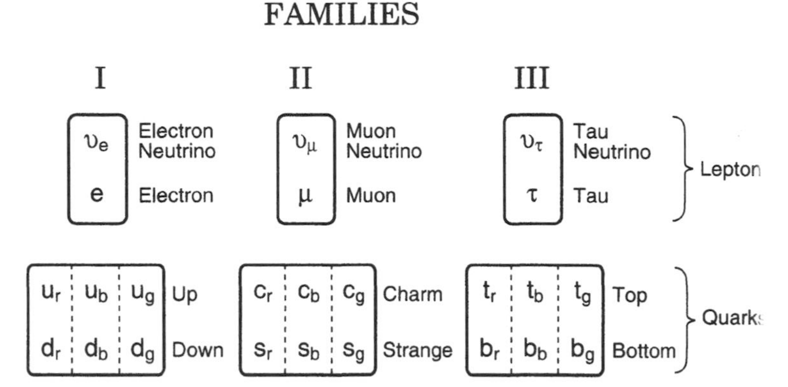


**FIGURE 4.1** *The matter particles, leptons and quarks. All of them have spin of magnitude 1/2 (in units we will ignore). The leptons come in six flavo (e, $\nu_e$, $\mu$, $\nu_\mu$, $\tau$, $\nu_\tau$), which are grouped into three families as shown. Withi each family they can be converted into one another, but not across familie (i.e., $e \leftrightarrow \nu_e$ can happen but not $e \leftrightarrow \nu_\mu$ or $\nu_e \leftrightarrow \mu$). The two members of eac family are called an electroweak doublet; they have electric charge differin by one unit, and different weak charge. Similarly, there are six flavors c quarks (d, u, s, c, b, t), grouped into three families. Again, within each famil conversions can occur, and the members have different electric and wea charges. For the quarks conversions can also occur across families but wit greatly reduced probability. Each type of quark comes with three differer color charges, r or g or b. The different color charge states of each type c quark are otherwise indistinguishable, since they have the same mass and th same electric and weak charges, so the electromagnetic and weak and grav tational forces cannot tell them apart. The names up and down, and top an bottom, were chosen for no deeper reason than their position in the doublet as shown.*

family structure exists. First, the electron, the muon, and the ta seem to differ from each other *only* in that they have differen mass. We can express all masses in terms of their ratio to th proton mass (rather than define a separate unit). That ratio i about 1/1836 for an electron, 1/9 for a muon, and 1.9 for a tau The formulas used to calculate predictions for every known prop erty and interaction of an electron depend on the value used fo the electron's mass. If the electron's mass is replaced by the muon's mass, the corresponding predictions for the muon are obtained; similarly for the tau. Every prediction for all three leptons has been confirmed experimentally, limited only by the precision of today's experiments. Thus nature seems to have replicated the electron twice (or, more objectively, perhaps it was the tau or the muon that was replicated twice).

Each of e, $\mu$, and $\tau$ has an associated particle called a neutrino, which carries no electric charge; $\mu$ and $\tau$ have the same electric charge as the electron. The Standard Theory also describes the interactions of the neutrinos correctly. A beam of electron neutrinos hitting a target will cause the production of electrons moving in a certain direction with a certain energy with a certain probability. Precisely the same statement holds for a beam of muon neutrinos producing muons. The equations that describe production of electrons also describe the production of muons if the muon mass is used instead of the electron mass. The masses of the neutrinos are not measured yet, but we know that they are very small or zero; for example, the electron neutrino mass would have been detected by now unless it were at least 50,000 times lighter than the electron. We will return to consider neutrino masses later; they are expected to be guides to how to extend the Standard Theory.

The second feature that tells us that there is a family structure is that transitions can occur that turn an electron into an electron neutrino (and vice versa), or a muon into a muon neutrino (and vice versa), or a tau into a tau neutrino (and vice versa)—but never an electron into a muon neutrino or tau neutrino, or a muon into an electron neutrino or tau neutrino, and so on. A clear line is drawn between the lepton families. The electrons occur in the atoms we are made of. The muon and the tau have very short lifetimes and are only produced at accelerators and by cosmic rays.

The spectrum of quarks is very similar to the leptons. There are six flavors of quarks: up (u), down (d), strange (s), charm (c),

bottom (b), and top (t), in order of increasing mass. The up
charm, and top quarks have electric charge that is $-2/3$ the elec
tric charge of an electron, and the down, strange, and botton
quarks have $+1/3$ the electric charge of an electron. The quark
also occur in three families that replicate each other's propertie
except for masses (see Table 4.1). The quark masses, again i
ratios to a proton mass, are about 1/235 for up, 1/135 for dowr
1/6 for strange, 1.6 for charm, 5.2 for bottom, and 170 for top
they have been measured only to about 20 percent accuracy a
present.

There are two major differences between quarks and lep
tons. First, quarks can make transitions within families just a
leptons can—from u to d or vice versa (u ↔ d), for example—bu
they can also make much weaker transitions to other quarks o
different electric charge, u ↔ s and u ↔ b (but not u ↔ c o
u ↔ t) etc. Second, quarks carry an entirely new kind of charg
that leptons do not carry, the color charge (introduced i
Chapter 1). Unlike electric charge, color charge does not occur i
the everyday world. All of its manifestations are indirect. It i
called "color" charge because the rules for combining quark
into protons and other hadrons are reminiscent of the rules fo
combining primary colors to get white light, but it has no actua
connection to real colors.

The Standard Theory tells us that no particle made of quark
and gluons, that is, no hadron (the composite particles includin
protons, neutrons, pions, etc. are collectively called hadrons) ca
carry any net color charge. That means that every process involv
ing quarks has to begin and end with hadrons, even though it i
the quarks we would like to observe. Imagine assigning th
quarks the primary colors, red and green and blue. The anti
quarks can also be used in building hadrons; they have the com
plementary colors. Then there are two and only two ways t
make "white" objects—that is, objects with no net "color" tha
could be hadrons: either combine all three different primary co
ors, or combine each primary color with its complement. Th

first gives the subset of hadrons called baryons, which includes the proton and neutron and others, while the second gives a second kind of hadrons, called mesons (pions, kaons, and others). One of the many achievements of the Standard Theory is to accommodate in a basic way the need for two and only two types of hadrons (baryons and mesons) since these are the two types that actually exist. And the attractiveness of calling the new and unfamiliar charge carried by quarks a "color" charge is apparent.

Particle physics in the 1950s and 1960s was largely focused on hadrons; quarks were unknown or barely introduced. After the early 1970s the focus changed to quarks and leptons. But quarks always appear in hadrons, so the use and study of hadrons is essential to get at quarks—it is with hadrons that experiments begin and end if quarks (or gluons) are involved. Our description of experiments will therefore often include hadrons as intermediaries.

An electron or proton or neutron can be separated from other particles and studied in isolation, but a quark cannot because it has to be bound in a hadron. That has led some people to question whether quarks exist in the same sense that electrons do. Careful analysis shows that the differences between quarks and other particles in this respect are only a matter of degree. Electrons are gravitationally bound to the earth, and electromagnetically bound to other charged particles, but very loosely because the gravitational and electrical forces are not strong. For an electron (or anything) to be seen or detected, photons must bounce off it and recoil into a detector (the eye, or a camera film, or an electronic physics detector). Quarks are bound these same ways but in addition very tightly bound by the strong force. To see a quark, photons (or some other particle) are bounced off the quark. In this case the projectiles must be energetic enough so that an eye or a camera can't detect them, but an appropriate detector can.

There is indirect but good evidence that no more families

like the three we know of exist. The evidence tells us they don' exist even if their quarks and charged leptons are heavier thar could have been directly produced at colliders today. If ther were new families they would affect measurements in two ways First, even though their quarks were heavy, their neutrinos ma be light like the other neutrinos. Then the Z boson would deca sometimes into the new neutrino plus antineutrino and tha decay would have been detected at the CERN LEP collider. Sec ond, recent measurements at the SLAC collider called SLC woul give one value for a certain quantity if a new heavy family di exist, and a slightly different value if a new heavy family did nc exist. The first reports provide evidence that additional familie do not exist.

## *Forces and the Particles That Transmit Them—the Bosons*

The Standard Theory accommodates the four forces. I sa "accommodates" because the theory does not without som further input require these forces or forbid additional ones; suc stronger conditions will, we hope, be part of an extended theor someday. Gravity is a kind of bystander because the Standar Theory is really about the weak, electromagnetic, and stron forces; the gravitational interactions of all the particles are muc much weaker than even their weak interactions.

After those of gravity, the effects of the electromagnetic forc are next most familiar to us. Any particle with electric charg sets up an electromagnetic field throughout space. The intensit of the field decreases further away from the particle (as th square of the distance). Any other particle with electric charg moving through the field set up by the initial particle will feel th effect of that field and change direction (like charges repel, oppc site charges attract). To change the direction of the moving pa ticle the field has to push or pull it, transfer some energy an momentum. Because the Standard Theory is a quantum theor the field energy and momentum are quantized—they come in chunks. For electromagnetism the packet of energy and momentum is called a photon. The light we see is composed of photons transferring energy from the outside world to atoms in our eyes. When two electrically charged particles have interacted via the electromagnetic force; we say that their interaction is mediated or transmitted by the exchange of a photon. Just as the words "force" and "interaction" can be used interchangeably, so can "transmitted" and "mediated." The photon, and any other particle that transmits a force because it is the quantum of a field, is called a "boson." The quantum theory of electromagnetism is called "quantum electrodynamics," sometimes abbreviated QED.

The electromagnetic interaction affects every particle with electric charge (the leptons e, $\mu$, $\tau$, all the quarks, and the W bosons described later in this chapter); neutrinos and the photons themselves and the gluons (described later in this chapter) do not feel the electromagnetic force since they have no electric charge. Atoms are made of electrons (negative electric charge) bound to nuclei (positive electric charge from the protons) by the electromagnetic force. Photons are the glue that binds the electron to the nuclei. Atoms have no net electric charge since the positive charge of the protons in the nucleus is exactly cancelled by the negative charge of the bound electrons, so naively one might think there would be no force to bind atoms into molecules and molecules into flowers, food, and tables. In fact, the electric field set up by an atom is indeed essentially zero far away because the atom is neutral, but near the atom at a given place the electric field is not quite zero because some electrons are nearer to that place than the nucleus is, and others farther (see Fig. 4.2). This residual electromagnetic field, often complicated to calculate, determines all of the properties of materials that we encounter in the everyday world. Apart from gravity, which holds us to the earth and describes falling, only electromagnetism—not strong interactions or weak interactions—determines all of

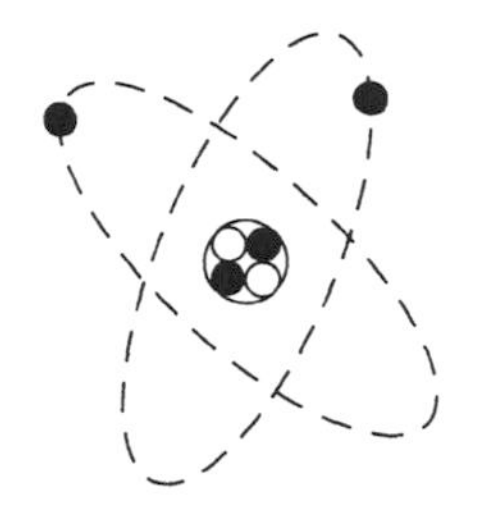

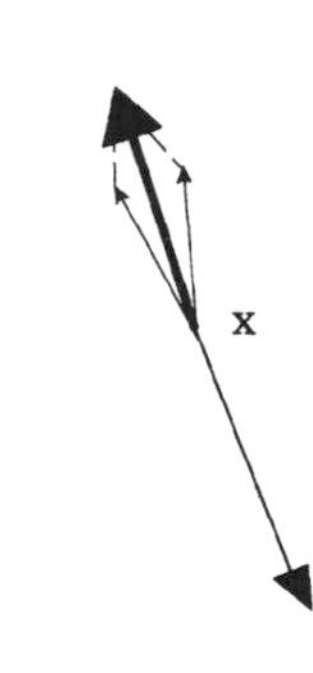

**FIGURE 4.2** *This is a drawing to illustra how the residual electromagnetic force outsic atoms gives rise to the force between atoms th leads to the formation of molecules. An atom i shown, with a nucleus having two protons an two neutrons. Two electrons are shown as do on dashed lines representing the electron "orbits." The electric fields at a given instant a the point x near the atom are shown; the field due to each electron are shown as thin arrow pointing toward each electron, and the net fie due to the two electrons is shown as a thic arrow. The length of each arrow is proportion to the strength of the corresponding field. Th field due to the two protons is shown as a th arrow pointing away from the protons. Becaus x is near the electrons and protons the thic arrow and the thin one, while nearly opposit are of noticeably different lengths and do n quite cancel. The small difference between the represents the net electric field due to the ato at that point. Another atom at that point wou feel that field and be attracted or repelled. the point y, further away, the fields are small (decreasing like the distance squared) and th arrows more parallel so the cancellation more complete and an atom there would feel much smaller force. A similar picture for th color force surrounding neutrons and proton accounts for the nuclear force.*

the diverse phenomena we encounter in our world: flowers, clothes, food, medicines, TV, pets, cars, Mozart, Picasso, rain, wind, sight, the human brain, mountains, rainbows, and so on. All the phenomena studied by scientists who do atomic physics, molecular physics, solid state physics, materials science, chemistry, and much more are due ultimately to the electromagnetic interaction.

The weak force has no easily recognizable effects in the everyday world, but it is nevertheless of great importance. Just as the photon mediates the electromagnetic force, the weak force is mediated by quanta of the weak field. The amount of weak charge that particles can carry comes with more possible values than electric charge, so three bosons, called $W^+$, $Z^0$, and $W^-$ (these are pronounced "W-plus," "Z-zero," and "W-minus") are required to transmit it. The weak force is actually weak not because its strength is less than the electromagnetic force, but because the range of the force is very short, so it is very unlikely that two particles are close enough together for one to feel the other's weak force. The range is very short because the W and Z bosons that mediate the weak force are so heavy that it is hard for two particles to exchange them. The weak field around a lepton or quark extends a much shorter distance than its electromagnetic field.

All particles with weak charge (all quarks and leptons, and also $W^+$, $Z^0$, and $W^-$) feel the weak force; only photons and gluons do not. The weak interaction is too weak to bind particles into "atoms." Rather, its most dramatic effect is that it makes all the quarks but the lightest one (u), and all the electrically charged leptons but the lightest one (e), unstable. Because of this instability the heavier quarks and leptons decay into the up quark and the electron and neutrinos; the muon and the strange quark decay in about one-millionth of a second, and the other quarks and leptons even faster. In a universe with no weak interaction many different kinds of atoms and nuclei would have existed, leading to many different phenomena.

The weak interaction also plays an essential role in the pro cesses that make the sun shine, thus providing the energy sourc essential for life on earth, and in building up heavy element essential for some of the chemistry necessary for our bodies t function. The first step in the production of energy in the sun i the process where two protons collide and produce a neutrin plus a positron and a deuteron (a nucleus that has one proto and one neutron). This process is a weak interaction, mediate by a W. Perhaps the best way to think of W's and Z's is as object like photons but very heavy, whose effects are crucial for the su to shine, and that cause most quarks and leptons to be unstable

The effects of a magnet and the static electricity we feel fron walking on a wool rug are very different, but it turned out tha the same underlying theory, Maxwell's equations of electromag netism, describes them both. A similar result holds in the Star dard Theory. Electromagnetism and the weak force seem ver different, but they become unified in the Standard Theory—th same underlying equations describe both. When probed at ver short distances with very energetic projectiles the weak and elec tromagnetic forces are no longer different. They seem differen at the low energies and large distances of our world because th quanta that mediate the weak interaction, $W^+$, $Z^0$, and $W^-$, ar very heavy and thus much more difficult to transmit betwee interacting particles than a photon.

The unification of the theory of the weak and the electromag netic interactions into the theory of the electroweak force is nc quite complete. On the one hand that seems to be a failing of th theory, but it also provides an opportunity. Physicists think of th situation in a rather abstract way that can be described along th following lines. In a more fundamental view of the world (partl described in Appendix B), there is a partial theory that describe some particle interactions and that is a lot like electromagnetisr but not quite the same, and another partial theory that describe some particle interactions and is a lot like the weak interactio but not quite the same. One can imagine the unification as a appropriate merging of these. Each partial theory has a particle like the photon, but nature has only one photon, so a compromise has to be worked out. The compromise is expressed in terms of a rotation in an abstract imaginary space, so that each of the candidates is rotated to point in a new direction in the imaginary space and they are combined to describe the actual photon of our world. The rotation is expressed in terms of an imagined "angle" called $\theta_W$ (Greek $\theta$, sub W, which we pronounce "theta-W"). It is called the electroweak unification angle; we will meet $\theta_W$ again in later chapters.

If the theory were truly unified, then it could predict the value of $\theta_W$. But the value is not predictable by the Standard Theory; rather, so far it has to be measured in experiments. That has been done, with the result that $\theta_W = 28.8°$. (To be precise, because all measurements have errors, the way physicists say this is that currently at 90 percent confidence level $\theta_W$ lies between 28.65° and 28.95°.)

Some people, unhappy at the inability of the Standard Theory to predict $\theta_W$, have said that the weak and electromagnetic interactions are not unified at all. But that is too strong a statement. $\theta_W$ enters into many predictions, and it has now been measured in about fifteen independent experiments, all of which give the numerical value just mentioned. If there were no unification, then what we called $\theta_W$ in one experiment would have been unrelated to what we called $\theta_W$ in another, so it could have been −90° (or anything) in one experiment and +42° (or anything) in another. That it always comes out the same no matter how it is measured is very strong evidence for the unification.

The opportunity that arises because $\theta_W$ is not determined in the Standard Theory is exciting to many particle physicists. In the search for how to extend the Standard Theory we do not have many experimental clues. Here is a major clue, giving us the opportunity to find principles in theories that extend the Standard Theory such that they correctly predict $\theta_W$. The value 28.8° does not look very simple or easy to get. But it turns out that a certain quantity, the square of sin($\theta_w$), occurs naturally i some theories, and has the value 3/8. Then it has to be modifie by certain quantum theory effects. After that is done, $\theta_w = 28.8$ indeed emerges. A certain class of theories that we will look at i Chapter 10 (the "supersymmetric grand unified" theories) suc cessfully predicts that value.

The strong force is mediated by the exchange of gluon quanta of the color field set up by any particles that carry colo charge (only the quarks and the gluons themselves carry th color charge—the leptons and $\gamma$, $W^+$, $Z^0$, $W^-$, do not). The wea force requires three bosons to transmit the weak interactio between particles of different weak charge. The color forc requires eight gluons, each with different color charge, to med ate all effects of the color force. "Color force" and "strong force are two different names for the same force; it is also calle "quantum chromodynamics," abbreviated QCD. The stron force binds quarks into protons and neutrons. As we saw above the protons and neutrons have no net color charge, just as atom have no net electric charge. But just as there is a leakage o electric field outside an atom that gives the force between mole cules (see Fig. 4.2), here there is a leakage of color force outsid the proton and neutron. This residual color force is the nuclea force that binds the protons and neutrons into the nuclei of th chemical elements.

All particles have another property, called "spin," a propert that behaves analogously to ordinary spin. The amount of spin particle has can be expressed in terms of a unit called Planck' constant, **h**, and because in a quantum theory things come i discrete chunks, particles can carry 0 spin, spin 1/2, spin 1, spi 3/2, etc. (the basic unit for spin is always $h/2\pi$, so the unit i usually omitted). For the basic particles of the Standard Theor only the first three of these occur. Higgs bosons have spin 0 quarks and leptons have spin 1/2; and the quanta of the electro magnetic and weak and color fields that mediate the forces ($\gamma$ $W^+$, $Z^0$, $W^-$, and gluons) all have spin 1. All particles whose spi is a half-integer multiple of the basic unit (here only quarks and leptons) are called fermions (named after the Italian physicist Enrico Fermi), and those whose spin is an integer multiple of the basic unit ($\gamma$, $W^+$, $Z^0$, $W^-$, gluons, h) are called bosons (named after the Indian physicist Satyaendra Nath Bose). Although having spin 1/2 or spin 1 may seem like a small difference, the properties and status of fermions and of bosons are very different, both from the point of view of constructing a theory, and also how they behave when more than one of them is present.

Now I can describe a beautiful and exciting aspect of the Standard Theory. Throughout the development of physics, learning the rules and learning the forces were independent processes. But slowly, as quantum theory became better understood, it was realized that the rules and the forces are connected. Today physicists understand that if we didn't know about any forces, but somehow we were to learn that some matter particle carrying some kind of charge exists (an electron carrying electric charge, say), then in order to have a consistent quantum theory two results must be true. First, it is necessary to introduce a new boson (in this case, the photon), which is the quantum of a new field (in this case the electromagnetic field), and second, the interactions mediated by that boson are a new force (in this case, the electromagnetic force). Thus, if electrons exist, then the theory requires that the photon must exist, and the form of the electromagnetic force is predicted. (A piece of jargon: The class of theories with the properties just named is called, for historical reasons, gauge theories.) Once we know that some particles carry a weak charge, then the existence of the $W^+$, $Z^0$, $W^-$ boson, and their properties, is predicted. Once we know that the quarks carry the new color charge, the existence of gluons, and their properties, is predicted. If we ever learn that the quarks or leptons or some new particle carries a new kind of charge, then we will know that new associated spin-one bosons exist.

The predictions of the existence of $W^+$, $Z^0$, $W^-$ and gluons, and their properties, were made during the period 1971–73. The gluons were detected, with exactly the right properties, at the German laboratory DESY, in Hamburg, in 1979. The $W^+$, $Z^0$, $W^-$ were detected, with exactly the right properties, at the European research center CERN in 1983 and 1984. The prediction and detection of these particles are remarkable intellectual achievements. Their prediction was based on pure reasoning constrained by a mathematical theory that embodied all the successful experiments and physical insights about nature from previous decades and centuries. In order to detect these particles, the predictions had to be trusted by people who could convince governments to spend hundreds of millions of (1977 dollars, and by hundreds of physicists who would spend almos a decade building the collider and detectors needed to confirm the predictions.

The quarks and leptons and charges we know about have been found in experiments. There is not yet a principle that tells us what other quarks or leptons or charges may exist (if any), so we must wait either for new experimental data or for such a principle to tell us that. To particle theorists the connection described here between the particles, the rules, and the forces is extremely exciting, hinting at still more powerful theoretical structures that will unite the particles, the forces, and the rules a a deeper level.

The four forces seem to be the minimal set in a sense, in contrast to the particles where we do not yet understand the need for families beyond the lightest one (though see Appendix C). Without gravity, stars and planets would not form Without electromagnetism, light and atoms would not exist Without the strong force, nuclei would not exist, so the sun would be an inert body that did not shine. Without the weak force processes essential for the sun to work the way it does would be absent, and heavy elements needed for life would not be made in supernova. Without any of the forces, life would not occur.

## *Renormalizability*

There are two more aspects of the Standard Theory to describe. One is the Higgs boson and the associated physics; I will defer that until Chapter 8 because it is an unfinished part of the Standard Theory and one that will point to the future as well as complete the formulation of the Standard Theory.

The other is an important conceptual and historical dimension. It has to do with the difficulty of formulating the rules. Understanding how to use a relativistic quantum field theory of electromagnetism (QED) was very hard. The theory was first formulated in the early 1930s, but initial attempts to use it to calculate some observable properties of atoms and particles gave answers that were infinitely large rather than meaningful numerical values. Slowly understanding improved, but the problem was not solved. Then, in 1947, an important number was finally measured (called the "Lamb shift" after Willis Lamb, who measured it); all attempts to calculate it had given infinite results before, but once it was finally measured the failure to calculate a meaningful value for it seemed much more serious. That quickly stimulated theorists to try improved approaches and, most important, convinced them to take the QED theory seriously and to calculate with renewed enthusiasm. Intense efforts in 1947 and 1948 succeeded, and allowed theorists to show that there was a systematic way to eliminate the infinite parts and identify the real numerical result. The procedure was called "renormalization." In the past, the need for the renormalization procedure has been criticized, but today the underlying physical reasons for the procedure are understood and theorists agree that the renormalization procedure is a sensible approach. In recent years it has been understood that renormalizations should occur in good theories, and that our understandings of different levels of scale in nature are related by renormalizations.

Those infinite answers prevented the formulation of good

theories of weak or strong interactions until the 1970s. Weal interaction theories were intrinsically more complicated thai quantum electrodynamics, and physicists were unable to detei mine if they were renormalizable. Finally in 1971, in Utrecht Netherlands, Gerard 't Hooft (then a student of Martinus Velt man) succeeded in proving that the Standard Theory was indee renormalizable. Knowing that the theory would allow the infin ties to be eliminated and the real predictions to be identified wa an essential step toward the full theory, as essential as its expei imental confirmations that are the subject of the next chapters.

Assuming the Higgs physics (Chapter 8) is eventuall resolved in a satisfactory way, the Standard Theory is then complete relativistic quantum field theory, the most mathemat cal and comprehensive theory that has ever existed. It embodie the rules of quantum theory and special relativity, and contain all of electromagnetism (Maxwell's equations). It describes a experiments on particles and their interactions. From 198 through 1993 several million experimental events were collecte at CERN in order to test predictions of the Standard Theory, an there is no significant disagreement with the theory. For the firs time in history there are no experimental inconsistencies or puz zles to solve in the domain of the theory. And now the domain o the theory extends up to the fastest speed (the speed of light out to the edge of the visible universe, and back to a billionth o a billionth of a second after the beginning of the universe. Th Standard Theory is here to stay.

# 5

# Experimental Facilities

The microscope and the telescope were the first devices that scientists used to significantly extend the range and power of human senses. Both were invented in the Netherlands, where optics was then most highly developed. In 1590 a Dutchman who made eyeglasses realized that since a convex lens (used to correct farsightedness) magnified, two might magnify more. He placed lenses at opposite ends of a tube, and had the first microscope. Steady improvements in lens grinding allowed better and better magnification; by the 1670s Anthony van Leeuwenhoek had achieved magnifications of 250–300. It took until 1608 for someone to notice that a convex and a concave lens appropriately arranged made a telescope. The discovery was made during Holland's war against Spain, and the telescope was immediately classified secret and put to use by the Dutch to give them an advantage in detecting advancing ships and troops. Despite Holland's efforts to keep the invention a secret, within a year Galileo, in Italy, heard rumors of the new instrument's existence. As with all aspects of science and technology, once an expert knows something is possible he or she can achieve it quickly (given the resources), and soon Galileo had a telescope that he turned to the "heavens." He quickly learned that the Milky Way was made of stars, saw four moons of Jupiter

moving in the "heavens," and much more. These advances ir
technology were crucial to the development of new scientifi
opportunities. In 1592 Galileo invented the thermometer, and a
little later the pendulum clock. Both were what we today cal
spin-offs, not invented for their own sake but because he needed
ways to measure temperature and time to in order pursue hi
basic research questions. From that time on, science and tech
nology have continually stimulated each other.

## *Colliders*

Four centuries of scientific and technological progress have le
to today's biggest "microscopes," the colliders of particle phys
ics, which enable physicists to "see" things a million millio
times smaller than what van Leeuwenhoek could see. When
probing so deeply, one cannot just look at what is there as one
would with a real microscope, because there is no substrate tha
stays permanently. Instead one has to concentrate a large
amount of energy in a very small region and cause a miniature
explosion, from which new particles will be created according to
the rules of quantum theory. Usually Einstein's equation E = mc
is used to emphasize the large amount of energy (*E*) that can be
made available by converting some mass (*m*) into energy, as in
nuclear power plant or a bomb. Particle physicists use it the
opposite way, putting energy in and converting some of tha
energy into new massive particles—creating particles that do no
normally exist in the universe (because they are unstable and
decay quickly, as described in Chapter 1). In particle physics thi
approach is so central that the units used to measure the masse
of particles are the equivalent amounts of energy needed to pro
duce them.

Getting that energy and concentrating it relies on two phys
ics principles. First, if a particle carrying electric charge is pu
in an electric field, it is accelerated and moves faster, so i
has more energy. Second, an electrically charged particle in a

magnetic field moves in a circle around the direction of the magnetic field.

Building a collider requires making a number of devices that set up an electric field. They get their energy the same way any household electrical device does—essentially by plugging into an outlet on the wall. The next step is to knock some electrons or protons loose from hydrogen, and put them into the electric field. The electrons or protons are then accelerated to higher energies, according to the first principle described in the preceding paragraph. Getting the electrons or protons to the desired high energies requires many devices in a row, since each device can add only a little energy. Magnets are used for this purpose, to bend the path of the particles into a circle, causing the particles to repeatedly traverse the path, gaining more energy each time. The faster a particle is moving, the harder it is for the magnets to bend it in a circular path, so eventually the particles cannot be kept in a circle if they gain any further energy—that sets the limit. For experiments that collide two beams, this whole process has to be done with two sets of particles, which are aimed at each other once they have been given the desired energy.

If the particles hit head-on they lose all their energy, concentrating a large amount of energy in a small collision region. That is an unstable situation, and the quantum theory tells us that the energy will quickly convert into a set of particles, each moving away from the collision region. Every single pointlike particle (quarks and leptons, W's and Z's, supersymmetric partners if they exist (see Chapter 10), and any others) has a certain probability of emerging from the collision if the energy equivalent of its mass is smaller than the available energy. A collider that accelerates its electrons or protons to higher energies than previous colliders will be able to produce both heavier and more particles.

In each collision some particles emerge. By now, most of them have been studied in previous experiments. The interesting ones occur only rarely, so lots of collisions are needed to produce them. The property of a collider that determines how many collisions will occur is called its "luminosity." In a real collide bunches of electrons or protons are accelerated together. Then a bunch moving in one direction is aimed at a bunch moving the opposite way, and it is hoped that a collision will occur as the bunches pass through each other. Each time the bunches pass through each other, typically only one of the many particles in each bunch collides, so the bunches can be made to produce collisions at several interaction regions, allowing a collider to have several detectors. The likelihood of a collision is larger if there are more particles per bunch, or if the size of a bunch can be made smaller (for a fixed number of particles), or if more bunches can be accelerated each second. The luminosity is more or less a measure of the ability of the collider to produce collisions, other things being equal. The two basic features of a collider are its maximum energy and its luminosity.

The field of physics called "accelerator physics" is devoted to improving our ability to accelerate particles. It has increasingly been recognized as essential for the future of research in particle physics. Considerable innovative work has been undertaken in the direction of achieving higher energies, though as yet less attention has been focused on luminosity.

Which particles are chosen to undergo collisions is largely a practical matter. Protons are easily available, relatively easy to accelerate, and good for achieving high luminosities. But the important interaction to study in order to learn new physics is the collisions of the pointlike quarks and gluons in one proton with the quarks and gluons in the other proton. Each proton is made up of a number of quarks and gluons, so in practice only a fraction of the energy of the proton (about a tenth or less) is carried by each quark or gluon that will actually collide. The tota available for making new particles is then at most abou 10 percent of the original energy given to the two protons.

That inefficiency has led to considerable effort aimed toward colliding electrons; because they are pointlike, all of their collision energy can be used to make new particles. For technica reasons the probability of making interesting new particles is larger if electrons and positrons (the antiparticles of electrons) are collided. This is not because of the energy from annihilating the electron and positron, but because when the electron and positron annihilate, the net electric charge is zero, whereas if two electrons collide, two of the final particles must carry electric charge (since the total amount of electric charge can never change) in addition to any other interesting particles created. But positrons are hard to get since any that are created soon annihilate on some electron, making it difficult to build an electron-positron collider of high luminosity. Since quarks of both electric charges are in every proton, this problem does not occur for proton collisions.

There is a second difficulty with using electrons. If any electrically charged particle is made to move in a curved path, there is a probability that it will emit photons that carry away some of its energy. That probability increases rapidly with the energy of the particle, and at the CERN LEP collider described later in this chapter that probability is so large that a noticeable part of the energy supplied goes into this radiation rather than into increasing the electron's energy. This is not a serious problem for protons because the relevant probability is smaller (by the fourth power of the masses) if a heavier particle is accelerated, and protons are nearly 2000 times heavier than electrons. But it does mean that the CERN LEP collider is the last circular electron collider that will be built to work at the frontier of physics. Future high-energy electron colliders will have to be linear ones, where the beams are accelerated in a straight line and aimed at each other.

To summarize: it is feasible to make proton colliders of very high energy and luminosity, but they are inefficient in that less than about 10 percent of the energy is used to make new particles. Electron-positron colliders use the energy much more efficiently, but it is much harder to make high-energy or high-luminosity ones. Proton colliders are better suited fo achieving many of today's physics goals, which mainly involv searching for new particles of unknown mass, but perhaps some day that will change. Later in this chapter I will briefly describ the colliders that exist or are planned.

## *Detectors*

To learn new physics at colliders, not only do collisions have t occur at sufficiently high energies and with sufficient frequenc the emerging particles must also be detected and the occurrenc of a new result, if there is one, recognized and recorded. Most o the heavier particles produced (W, Z, c, b, t, $\tau$, and any as ye undetected ones) are unstable, with very short lifetimes, so the travel far too short a distance to be directly detected. Instead their existence must be deduced from the behavior of the parti cles they decay into. For example, electrons and positrons fre quently emerge from collisions. In general they can have almos any energy. But if they come from, say, the decay of a Z boso into an electron plus a positron, then their total energy has t equal the energy equivalent of the Z mass. So the energies of a electrons and positrons in a given event are checked for thi equivalence, and if it holds then that electron and positron ar replaced by a Z boson in the analysis of that event. This kind o analysis is repeated for all the particles in the event, an repeated each time such an identification is made, until the even cannot be simplified further. Then the number of each type o event is checked against the number predicted by the Standar Theory. Something new might be recognized because a large number of events of a particular type occurred than was pre dicted by the Standard Theory. The characteristics by which new particle can be detected are called its "signature." One sig nature of a top quark is events with a muon, an electron, two jet of hadrons (arising from quarks), and some missing energy; th probability of seeing such events caused by anything other tha a top quark is very low, so finding several such events at Fermilab is one way to know that a top quark is being detected. The collider at Fermilab is the only one with sufficient energy to produce top quarks until the CERN "Large Hadron Collider" is constructed, perhaps by 2005.

Of all the particles, only electrons (e), muons ($\mu$), photons ($\gamma$), neutrinos ($\nu$), and hadrons emerge into the detector. A detector is basically made of several layers, each of which responds differently to at least one of these particles, so that after traversing all the layers the particle type is uniquely specified (see Figs. 5.1 and 5.2). Then the particle energy is measured by one of two techniques—either it loses its energy inside the detector by colliding with detector particles, in which case that energy effectively heats the detector a little and the amount of heat is measured, or the tracking chamber is built inside a magnet and the particle then curves around the lines of magnetic field. Since the amount of curvature changes with the particle's energy, the energy can be measured.

Two special cases occur. Quarks do not appear as a single charged particle, but as a narrow "jet" of hadrons, for reasons explained further in the next chapter. And neutrinos have only weak interactions, so weak that they do not interact in the collider detector; once it is created in the collision or a decay, a neutrino just escapes the collider detector. Luckily it can still be "detected" because it carries away energy—if the energies of all the particles in the event are added up the result should equal the original amount, and if it does not, it is assumed in the analysis of the event that the missing energy was carried off by a neutrino. If the resulting event type is one predicted by the Standard Theory that is OK—if not it would be a signal of something new.

There are a number of constraints that make building a good detector a difficult and challenging task. The dominant one is the rarity of interesting events, mentioned earlier. A new experiment

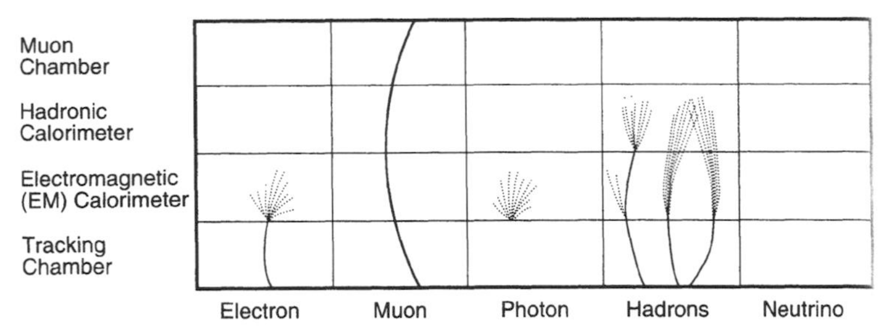


**FIGURE 5.1** *Shown schematically is the way a detector can be made b combining several layers, each of which responds differently to one or mor types of particles. It is assumed that the entire detector is in a magnetic fiel so electrically charged particles curve. The tracking chamber is nearest to th interaction, and the muon chamber furthest. In a tracking chamber ever charged particle leaves a track by some interaction with the atoms of th detector; for example, it could knock loose electrons from atoms it hit, an those electrons could then travel down wires and be counted at the edge. A the tracks in an event should point back to a single point if they all came fror one collision. An electromagnetic (EM) calorimeter is made of materials cho sen so that any electrically charged particle loses energy, more easily if it i lighter, so electrons typically lose most of their energy there, and hadrons los some. Photons also interact easily with the electrons in the EM calorimete and lose their energy. A photon can be distinguished from an electron becaus the photon only shows up in the EM calorimeter, while the electron shows u both in that and in the tracking chamber. Hadrons (pions, protons, etc.) shov up in the tracking chamber if they are electrically charged, then lose som energy in the EM calorimeter, and the rest in the hadronic calorimeter. Muon leave a track in the tracking chamber. They interact electromagnetically bu are about 200 times heavier than electrons so they lose energy less easily an almost all muons get through the calorimeters without interacting. Any par ticle that makes it all the way to the muon chambers (really just another laye of tracking chambers) is likely to be a muon. Finally, neutrinos do not shov up directly in any detector element, but carry away energy (as described i the text). Each of these detector elements should be thought of as a spheric layer surrounding the collision point.*

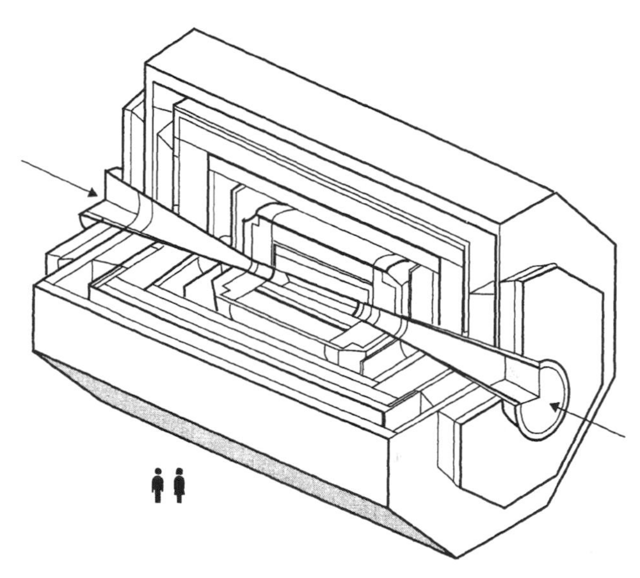

**FIGURE 5.2** *A schematic of a possible detector for a future collider such as the CERN LHC. Two people in the foreground illustrate the scale. The different elements correspond to the ones shown in Fig. 5.1 (though not in a simple way). The arrows show the beams that will collide at the center of the detector.*

may be at a higher energy than previous ones, so some process for which there was insufficient energy before can now occur, or a new experiment may have higher luminosity, so some process might occur that was too rare to see before. The goal is to recognize a few unexpected events, even though the actual particles that emerge into the detector are just electrons, muons, photons, neutrinos, and jets of hadrons. At a proton collider the initial "guess" that the event might be interesting has to occur on a time scale of the order of a millionth of a second, so it has to be made by a computer. A second decision, whether to save a record of the event or not, has to be made in perhaps a thousandth of a second—again by a computer. Further, every measurement has some error, and if errors accumulate in the full analysis

needed to interpret an event, it may not be possible to distin
guish between two interpretations, so the accuracy of measure
ments is crucial. Deep and extensive knowledge of electronics
computing, materials science, engineering techniques, and phys
ics goals is needed to make a detector that works. Technologica
frontiers are pushed further during the development of ever
particle physics detector.

## *Laboratories and Their Colliders and Detectors*

Because of their cost and complexity, there are only a few parti
cle physics laboratories in the world. The accelerators and co
liders at each one have evolved over time, and have had severa
detectors. Each detector today is so complex that hundreds o
physicists, engineers, and technicians are needed to build it
Experimental results and discoveries are thus associated witl
detectors rather than with individuals, though one or a few peo
ple may have been essential to get the collider or detector buil
and functioning as it does. In order to be able to refer to then
later, and to inform the reader about the labs where experiment
are done, I will summarize the main facilities existing today
Important measurements and discoveries have been made at al
of the labs. My purpose here is not to provide a history, but only
to gather in one place a description of the main facilities. I refe
to them throughout the book, particularly in the next chapte
where experiments done at these labs are described.

For this summary we need units in which we can express th
particle energies—perhaps the simplest choice is to use electron
volts, abbreviated eV. The numbers can get very large, so prefixe
are used: M for million (MeV), G for giga meaning a billion (GeV)
and T for tera, a million million (1 TeV = $10^{12}$ eV). "Giga" is in
"gigantic;" "tera" appears in some words, especially words used
in biology, meaning "monstrous."

## SLAC AND SLC

Perhaps the most successful laboratory from the point of view of the Standard Theory has been the Stanford Linear Accelerator Laboratory (SLAC), in Palo Alto, California. Several of the major discoveries that form the foundation of the Standard Theory were made there. A two-mile-long electron linear accelerator was constructed at SLAC in the early 1960s; it could accelerate electrons to 17 GeV. When those electrons were used to probe protons deeply, quarks were found (Chapter 6 contains brief descriptions of a number of the basic experiments). Beginning in the early 1970s a small circular electron-positron collider called SPEAR (whose total energy could range from about 2.5 GeV to about 5 GeV) took data at SLAC, discovering the charm quark, the tau lepton, and much more. (Although SPEAR has since been closed for particle physics, it is now being used as a "synchrotron light source," basically an intensely energetic x-ray source to study materials and biological systems. This important area of physics was a spin-off from the accelerators developed for particle physics.) SPEAR had one detector, called MARK I.

To avoid the problem that electrons and positrons moving in a circle lose energy, efforts began in the early 1980s to build a linear electron collider, based on a clever idea developed by Burton Richter. By appropriate use of electronics and new technology both electrons and positrons are accelerated in the existing linear accelerator at SLAC, now upgraded to 50 GeV, side by side. Then at the end electrons are bent in a half circle one way and positrons the other, so they can meet and collide head-on. Since they are only bent for one half circle instead of many circles, their energy loss is minimal. This facility, called SLC, can reach a total energy of 100 GeV. It was designed both to be a research and development device for possible future linear colliders, and to do experiments. It has been very successful for the

**FIGURE 5.3** *End view of the SLD detector at SLC. The beams are perpendicular to the page, and collide inside the detector. Photograph courtesy of DOE/SLAC.*

former purpose; it should be thought of as the first linear collider, to be followed by bigger and better ones. Because achieving high luminosity turned out to be very difficult, and partly because the project was attempted with a minimal budget, SLC only began to do significant particle physics in 1993; it should make some important measurements in the mid-1990s. SLC has one large detector, called SLD (Fig. 5.3). In 1993 SLAC was approved to construct a very-high-luminosity "b quark factory" to study the effect called CP violation (see Appendix C). The b-factory will begin to take data as SLC is completing its physics program, in the late 1990s.

### BROOKHAVEN

The first large accelerator constructed after World War II was built at Brookhaven National Laboratory, on Long Island, New York. It accelerates protons to 32 GeV and then hits fixed targets with them, producing secondary beams of neutrinos (described in the next section) and other particles that can then be observed and studied. The charm quark was independently discovered there, as were a number of important phenomena that led to the Standard Theory.

### FERMILAB

The most energetic facility in the world today is the proton-antiproton collider at Fermi National Accelerator Laboratory (Fermilab) just west of Chicago. Its proton and antiproton beams each have energies of 1 TeV. It is arranged so that collisions occur at two intersection regions. The two detectors at these intersections are called CDF and D0 (D-zero), each built and operated by a group of several hundred physicists. The highest priority at Fermilab is directly detecting the top quark and measuring its mass and other properties. The Fermilab collider is also a good place to search for new heavier particles such as supersymmetric partners (Chapter 9) that we hope will tell us how to extend the

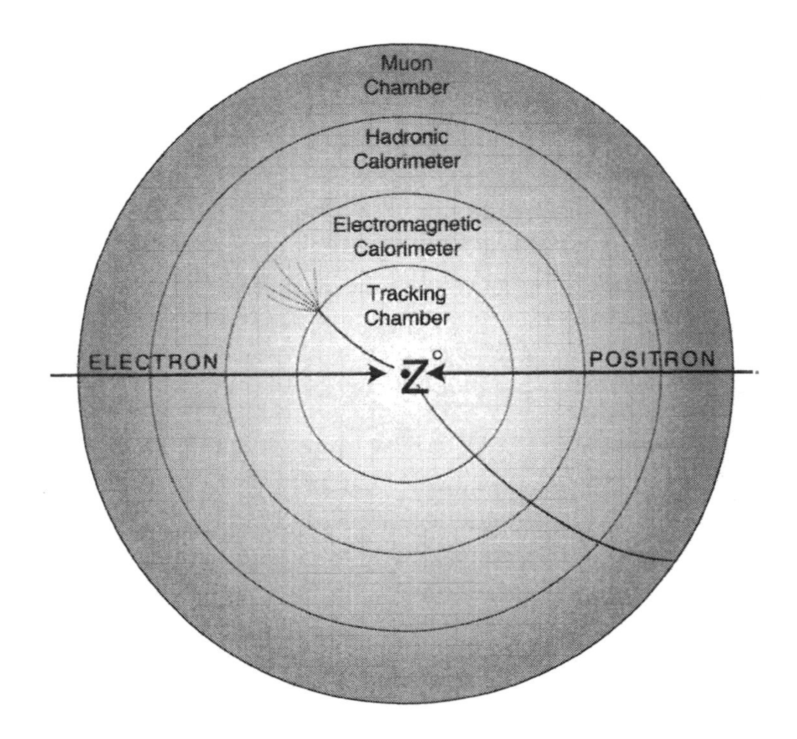


Standard Theory. It is possible to increase the luminosity of th Fermilab collider quickly and cheaply by a factor of perhaps 10( which would make it a powerful facility to search for a Higg boson, and greatly extend its capability to search for ne\ particles.

## CERN: LEP AND LHC

The main European particle physics lab is CERN, located i Geneva, Switzerland. Like SLAC, CERN has repeatedly upgrade and diversified its facilities over three decades. The main on now is LEP, a circular electron-positron collider with initial tot energy up to 100 GeV that has made many measurements c unprecedented accuracy in order to test the Standard Theor

**FIGURE 5.4** *This diagram illustrates how an event might look in a detector and how it is analyzed. The (spherical) layers are the elements of the detector, as in Fig. 5.1. An electron and a positron collide and produce a $Z^0$. That could be detected as follows. Two tracks occur in the tracking chamber. One is an electron and the other is a muon according to the criteria of Fig. 5.1. After analysis, it turns out that the total energy of the electron and the muon do not add up to what it would be if they came from a $Z^0$ decay, and electron plus muon is not one of the allowed $Z^0$ decays, so the event is not initially identified as a $Z^0$ decay. However, a process that the Standard Theory predicts should occur is $Z^0$ decaying to a tau lepton plus an antitau lepton, followed almost immediately by the tau decaying to an electron plus a neutrino plus an antineutrino, and the antitau decaying to an antimuon plus a neutrino plus an antineutrino. Only the muon from one tau and the electron from the other tau are detected; the taus themselves decayed before they travelled even a millimeter, too short a distance to be directly observed. Since the neutrinos escape detection, that hypothesis could describe this event, so the hypothesis is tried, and works. If no Standard Theory process accounts for an event, new kinds of ideas are tried. Each new kind of idea suggests other kinds of events that might occur, and they are searched for. It the tau lepton had not been previously detected it could have been discovered this way, as a new decay of $Z^0$. Which track is the muon and which the electron can be seen by comparison with Fig. 5.1.*

The total energy of LEP is being increased to 180 GeV, and perhaps eventually higher. LEP has four collision locations, each with a large detector (whose names are ALEPH, DELPHI, L3, and OPAL). CERN plans to build a large proton-proton collider, LHC, perhaps to begin taking data in 2005; until installation of LHC begins, LEP will accumulate data. LHC will have a total energy almost seven times that of Fermilab, and a luminosity over fifty times that of Fermilab's present luminosity. A view of CERN and LEP and its detectors is shown in Fig. 5.5.

### DESY: PETRA AND HERA

The next largest lab in Europe is the German DESY, in Hamburg. Its second collider, PETRA, first detected gluons and confirmed many of the predictions about quark jets. Today it has a higher energy electron-proton collider called HERA, which collides 800

GeV protons with 30 GeV electrons, particularly to study hov
quarks and gluons are bound to make protons.

### OTHER COLLIDERS

Several other particle physics labs are located around the world
KEK with the collider TRISTAN, and a b quark factory under con
struction, at Tsukuba City in Japan; an electron collider at Novo
sibirsk and a proton collider under construction at Serpukhov i
Russia; a small electron collider in Italy; an improved version o
the SLAC SPEAR collider in Beijing, BEPC, that produces a larg
number of charm quarks and tau leptons and has made a
important measurement of the mass of the tau lepton; and a
electron collider, CESR, that produces large numbers of b quark
and has made many valuable measurements of their properties
at Cornell University in Ithaca, New York. All of these facilitie
have made and are making important measurements of quar
and lepton properties, and have performed significant tests o
the Standard Theory.

## *Low-Energy and Nonaccelerator Experiments*

This book emphasizes the role of colliders in establishing an
confirming the Standard Theory. That is appropriate because th
energies accessible to the colliders of the past two decades wer
designed to be in the natural region to search for the W's, Z
quark and gluon jets, etc.—the Standard Theory particles. When
in later chapters, we emphasize completing and extending th
Standard Theory, colliders will still play a crucial role, but there i
another set of experiments that is also essential. The collide
experiments are like using a projectile to find out if a peach ha
a pit; the nonaccelerator approach is a little like waiting for th
peach to rot away to see if there is a pit inside, and looking at lot
of peaches in case some rot quicker. Positive results from any o
the following experiments will immediately take us beyond th

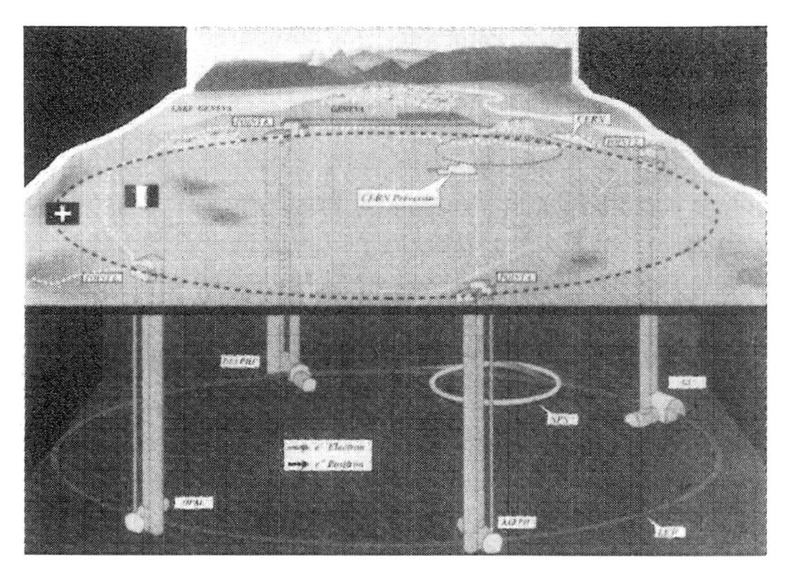


**FIGURE 5.5** *A schematic of the CERN laboratory, showing the Geneva airport, the city of Geneva, and the Alps in the background, and the LEP collider ring. The four detectors are about 100 meters underground. Part of the ring lies in France and part in Switzerland. Photograph courtesy of CERN.*

Standard Theory, and point the way to how to extend the Standard Theory.

## DETECTING PROTON DECAY

Theories that unify quarks and leptons (Chapter 11) allow the quarks in a proton to turn into leptons, so the proton can decay. Searches in the 1980s for the decay of protons could have observed decays according to the predictions of some theories, but did not; more sensitive experiments are now underway, by a mainly European collaboration in Italy, and a mainly Japanese experiment, with American and other collaborators, in Japan.

These experiments are remarkable accomplishments. The first estimates of how long a typical proton might live before it decays was about $10^{30}$ years, very long compared to the lifetim of the universe ($10^{10}$ years). So there is no need to worry abou the protons in us decaying, and no point in staring at a particula proton and waiting for it to decay. But quantum theory tells u that unstable particles can decay at any time, so if over $10^3$ protons could be put in a container, a few of them are likely t decay each day. That was done by putting a tank (20 meters o each side) deep underground in a salt mine (outside of Cleve land) in order to have the ground absorb the cosmic rays tha could mimic a proton decay, and filling the tank with water. Sinc the proton might decay (e.g., to a positron and two photons anywhere in the tank, the water had to be clear enough so on photon could travel through meters of water to a detector on th side of the tank. As one of the experimenters once remarked, i you filled the Michigan football stadium with water that clea you could easily watch a game from the top of the stadium. (Afte considerable study of sophisticated methods to purify water, th experimenters found that some Culligan equipment would d the job perfectly well. They called the national Culligan head quarters, and learned that the only available unit to do the jol happened to be in Ann Arbor, Michigan—which was right wher they had wanted to do tests.) Unfortunately, the initial experi ments (the one near Cleveland, a similar one in Japan, anothe with a different detector approach in Minnesota, and one i Europe) did not see protons decay. The significance of whethe or not protons decay is large enough that a new generation o experiments is underway, one in Japan and one in Italy, bot eventually perhaps a hundred times more sensitive than th first ones.

### MEASURING NEUTRINO MASS

Although in the Standard Theory neutrinos have zero mass no principle requires that, and extensions of the Standard The ory almost always imply that neutrinos have small masses Experiments have been attempted for over a decade to try t measure neutrino masses, and several more sensitive ones are underway. Some international collaborations (with underground detectors in Canada, Russia, Italy, and Japan) have important experiments attempting to detect neutrinos from the sun—the behavior of the solar neutrinos depends on their masses.

These detectors are again placed deep underground so no cosmic rays cause an apparent event that mimics the behavior of a signal event. The solar neutrinos, like all neutrinos, interact very little (which is why they can escape from the sun), but a big enough detector can be built to expect about one solar neutrino to interact in the detector every few days. The first such experiment began in the late 1960s and has continued. It was a conceptual and technical tour de force, carried out by Ray Davis and collaborators. They knew that neutrinos from the sun should be absorbed by a chlorine nucleus, giving an electron and an argon nucleus. They figured out that in order to get about one absorption a week, about 100,000 gallons of liquid containing chlorine had to be stored in a tank, which was put in the Homestake Gold Mine, in Lead, South Dakota. Fortunately, off-the-shelf cleaning fluid is mainly chlorine, so filling the tank was not prohibitively expensive. Every few months the tank was emptied and about a dozen (!) argon atoms were extracted from the $10^{30}$ or so atoms in the tank, a truly amazing feat. To convince themselves (and the world) that they could indeed do what they claimed, the experimenters did "calibration tests," by sometimes introducing a known number of argon atoms and then finding them. This experiment did successfully detect solar neutrinos for the first time, but the number detected was only about half what was expected. Data from other related experiments has increasingly convinced physicists that this deficit is real, and that the cause may be a subtle effect that can occur only if neutrinos have mass. The question is not yet settled, and several experiments are underway to try to clear it up, and to measure the neutrino masses if they are indeed the cause of the deficit.

At CERN, Brookhaven, Fermilab, and Los Alamos beams of

neutrinos created in collisions of the proton beam are being stud
ied using a different technique to learn whether neutrinos have
mass. To make neutrino beams, protons are shot into a piece of
metal. Lots of mesons, baryons, photons, muons, and neutrinos
emerge. Magnets are placed behind the target, and they bend
the path of any particle with electric charge out of a straight line
Some material is put in the path to absorb all other particles
(photons, neutrons, etc.). After a few feet, only neutrinos can get
through, because they interact too weakly to be absorbed. Then
very large detectors are placed in the path to study the neutrino
interactions, with the hope of gathering a few events a month
The number of events observed, and some detailed properties of
the events, will provide information about neutrino masses
These are important and difficult experiments.

### SEARCHING FOR FORBIDDEN DECAYS

In the Standard Theory a list can be made of all simple interac
tions and processes that the quarks and leptons and hadrons can
undergo. (The way this is done in practice is described in Appen
dix A.) When that is done, some simple processes whose occur
rence could be imagined without the theory are not among those
expected to occur. These processes are described as being "for
bidden" by the Standard Theory. One forbidden process is the
decay $\mu \rightarrow e + \gamma$. If this process had occurred at the same ("big")
rate as normal muon decays, then the Standard Theory would
have been wrong; if it occurs at a tiny rate, it will give us a needed
clue about extending the Standard Theory. Other forbidden proc
esses are kaon $\rightarrow \mu + e$, $\tau \rightarrow e + e + \bar{e}$, and related ones
Experiments at a number of laboratories around the world are
searching for such decays. If any of them are detected it will have
a major impact on the development of particle physics.

For the past century, until the 1970s, experiment led theory
in basic physics. (The exception was the area of quantum theory
predictions after 1925, when many theoretical predictions were
tested, confirming that the quantum theory did indeed provide

the rules to describe nature at the scale of atoms.) Since 1970 in particle physics theory has again been ahead, with every discovery predicted except possibly one (the tau lepton). Today when the emphasis is on completing and extending the well-confirmed Standard Theory, a number of major predictions could be tested, and a number of major discoveries could occur. Unfortunately, the facilities required to make those tests and discoveries have become large and expensive; the need for bigger "microscopes" is unavoidable. At the same time, society's commitment of recent decades to very basic research is increasingly in doubt, particularly in the United States.

# 6

# The Experimental Foundations of the Standard Theory

Some clever person coined the modern proverb "Theory without experiment is like a bird withou legs." This chapter describes the many legs on which the Star dard Theory stands. Almost all of the experiments that will b described here would be on every particle physicist's list of th most important ones. There is less agreement about which one to leave out, because some experiments played a major role i determining the way the field developed, but once the insight gained from them became incorporated into the theory, the looked less dramatic from today's perspective. Others were cru cial when they occurred but were superseded by later, mor direct ones. I have focused on those experiments that provid the most explicit confirmations of predictions or explanations c phenomena. A few have been mentioned in earlier chapters, bu for completeness I mention them here also; some are a singl experiment, and others the integrated effort of several relate experiments.

Often experiments are described in association with th names of one or a few experimenters. But almost all the exper ments that established the Standard Theory had crucial contri butions from a large number of people. It is significant that thes profound discoveries were made by groups of people working together, so I have not emphasized names or anecdotes.

### HADRONS AND THEIR PROPERTIES

In Chapter 2 I described briefly how, through the 1950s and 1960s, a large number of hadrons were discovered—in addition to protons and neutrons and pions, others called kaons, lambdas, etc., each having some property different from the rest, were found. Over two hundred hadrons are now known. Once hadrons are assumed to be made of quarks, theorists can predict which types of hadrons should exist and which should not. In addition, the spin and other properties of each hadron can be predicted. After forty years of work in this area there is a remarkable coherence of results, in spite of the difficulty created by the fact that detailed calculations are often not possible because technically complicated mathematical problems arise. To summarize the situation: (1) no hadrons have been predicted to exist that have not been found, (2) no hadrons have been found that have not been predicted to exist, and (3) all properties of hadrons (hundreds of measurements) are qualitatively and sometimes quantitatively understood.

Perhaps the most remarkable result of interpreting hadrons as objects made of quarks was that the color theory predicted that two *different* kinds of hadrons could be made of quarks: "baryons" (including protons and neutrons) composed of three quarks, and "mesons" composed of a quark plus an antiquark, and these two kinds are indeed what exist. Some additional hadrons (called "glueballs") are expected to be made of only gluons, no quarks, but they are thought to be rather heavy and it is not yet settled whether some experimental candidates for glueballs are really the predicted states. Altogether, after forty years of study of hadrons, the key point from our perspective is that there is no hint of any puzzle involving hadron properties that suggests the basic theory is not right, and there is much evidence from the hadrons that confirms the Standard Theory.

## QUARKS INSIDE PROTONS

In 1911 Rutherford scattered energetic projectiles from atom and discovered the existence of the nucleus of the atom becaus the projectiles bounced off at large angles. It took until 1968 t repeat that process with protons as the target. The result was th discovery of the pointlike quarks in the proton (in a SLAC exper iment, already described in Chapter 2). It is important to under stand that in both cases it was not only the existence c surprising recoils of the projectiles at large angles that suggeste that there was a pointlike object inside, but also that the numbe of recoils at each large angle could be calculated by assuming nucleus in the atom, and three pointlike quarks in the proton.

Although these experiments did not also directly demon strate the existence of gluons, they did show that the projectile interacted with only about half of the stuff making up a proton That is what would be expected if the projectiles were electron and neutrinos that carry no color charge, and if about half of th material in a proton consisted of gluons (which only interact wit particles carrying color charge), there to bind the quark together.

## THE STRONG FORCE IS LESS STRONG WHEN QUARKS ARE CLOSE TOGETHER

One puzzling result had been observed in the 1968 SLAC exper iments that detected the quarks inside the proton. The quantita tive prediction just mentioned, which implied the existence c pointlike quarks in protons, should (for technical reasons) hav been valid only when the projectile energy was large compare to the typical energy of motion of a quark in a proton. But th prediction turned out to be valid even for projectiles that wer not very energetic. At the time this fact made physicists nervou because if any aspect of a prediction is wrong it can indicate tha apparent agreement elsewhere is coincidental. When the theor of the color force was worked out quantitatively in the earl

1970s it was found, remarkably, that the color force behaved very differently from the electromagnetic and gravitational forces. The color force behaved somewhat as if the quarks were attached to the ends of a rubber band. The force between them increased rather than decreasing with distance as the other forces do. Under normal conditions the rubber band was not stretched, and the quarks behaved as pointlike free particles, just as had been found experimentally. (The jargon for this effect is "asymptotic freedom.") But once the rubber band was stretched it pulled back strongly and kept the quarks bound in a hadron, as had also been observed. The puzzle disappeared, and confidence in the Standard Theory grew.

### QUARKS APPEAR AS JETS

The theory of the color force makes a number of unexpected predictions. One of them is that if a quark is struck hard in a collision and separated from the quarks it is bound to, it does not emerge as a single particle but as a narrow "jet" of hadrons. (Some jets are shown in Fig. 6.2.) The reasons are rather subtle. Basically, the quark has to end up in a color-neutral hadron. The simplest way to do that is to find an antiquark and bind with it into a meson. The energetic collision has led to a number of quark-plus-antiquark pairs being temporarily produced. The struck quark can capture one of the antiquarks, leaving a meson (typically a pion because it is lightest) and the other quark of the pair, which is then pulled along by the strong color force and which then repeats the process. The theory predicts that the most likely situation is one where the energy of the original quark ends up shared among a number of hadrons rather than ending up mainly in one or two. Such jets began to be observed in the 1970s as soon as sufficiently energetic collisions were possible at larger colliders.

As with other things we discuss, it is not only the observation of a qualitative phenomenon that is convincing—there is also a quantitative aspect that is a powerful test. In this case there is a prediction for the direction that the quark jet has as it emerges from the collision. If the quark jets are produced by colliding an electron and a positron, then after many events are accumulated the number of jets produced perpendicular to the direction of the initial electron will be half the number produced in the direction of the electron, with a (known) smooth rate of change between those two directions. (This is a typical prediction in a quantum theory. It says that for any particular collision the jets will go in some particular direction. After many collisions, half as many should accumulate in the perpendicular direction as in the direction of the electron, 3/4 as many at 45° to the electron direction, etc.) This prediction depends on the spin of the quark being 1/2, so the entire jet of hadrons is behaving as if it has the expected spin of a single quark, a result that was surprising and beautiful even to the physicists who expected it. If the jets had not come from a particle of definite spin, the variation with angle could have had any form; for example, perhaps jets could have been produced with equal numbers at every angle. If the prediction had been for a spin-zero particle (instead of spin 1/2), it would have required the number perpendicular to the electron direction to be largest, rather than the smallest, decreasing to half as many at 45°, and giving no jets in the direction of the electron, very different from the predictions for a spin-1/2 quark. These results confirming the Standard Theory were first seen at SLAC in the 1970s, when its electron-positron collider called SPEAR began to operate. The results became less convincing when it was determined that the SPEAR energy was too low, and when a new higher energy collider, called PETRA (located at DESY in Hamburg) began releasing conflicting findings.

## THE QUARK PRODUCTION RATE MEASURES ITS SPIN AND ELECTRIC AND COLOR CHARGE

Before the mid-1970s, the results of electron-positron collisions seemed to elude interpretation. As the energy increased so did the number of hadrons produced, but not in any way that seemed interesting. When the data were reinterpreted in terms of production of quarks, followed by quarks forming hadrons, a simple and powerful picture emerged, one that led to confirmation of several basic Standard Theory predictions (at the SLAC collider). In general, particles with more charge interact more strongly, so there is a greater probability of producing them. One of the surprising properties quarks were predicted to have was an electric charge different in magnitude from those of electrons or protons—since a proton has one unit of electric charge, making a proton of two up quarks and one down quark would work if the up quark had electric charge $+2/3$, and a down quark $-1/3$ (in units of the proton charge). Then the second family, charm and strange (see inside back cover or Table 4.1), had to have the same assignments, $+2/3$ for charm and $-1/3$ for strange.

When electrons and positrons collide they lead to the production of each quark and lepton in definite amounts. The amount of each quark (u, d, s, etc.) expected is known once its spin, electric charge, and color charge are specified. If the particles produced are heavy, some of the energy of the colliding particles is transformed into the mass of the produced particles, so heavier quarks or leptons cannot be produced until the colliding particles have sufficient energy.

To work out the prediction we can take the probability of producing the muon as the unit. All the quarks and leptons have the same spin as the muon, so the same number of each of them would be produced if no other property mattered. The rules tell us that the number produced is proportional to the square of the electric charge. Also, there are three colors of each quark, all of which will be produced, but only one muon. Then the rule is: To calculate the number of quarks of a given kind produced relative to the number of muons produced, simply square the electric charge of the quark and multiply by 3 for the three colors. So the down and strange quarks each give $3(1/3)^2 = 1/3$ of the number of muons. The up and charm quark give $3(2/3)^2 = 4/3$ of the number of muons. Then when there is just enough energy to produce up and down and strange quarks, the rate for quarks should be 4/3 for up plus 1/3 for down plus 1/3 for strange times than the muon rate, which adds to twice the muon rate. After the additional energy needed to produce the heavier charm quark is available, the rate should be 10/3 times the muon rate, which is indeed the rate observed.

As this analysis shows, many parts of the Standard Theory enter into predicting the rates, and were confirmed by the results. Now such measurements have been done to much higher energies. They show the tau lepton and the b quark as expected, and nothing else, which implies that no other particles exist up to the largest energy so far available. At higher energies we expect the top quark to be produced too.

### DISCOVERY OF CHARM QUARKS

The direct discovery of hadrons made of charm quarks (Fig. 6.1) occurred in a dramatic way in November 1974. The discovery was dramatic not only in the form the data took, but also because it was discovered independently by two groups (see Chapter 2), one at SLAC and one at Brookhaven. There was no need to wait for confirmation. Until then, though the evidence for quarks and the Standard Theory had been mounting, only a few physicists were totally confident of the whole picture, and many were still skeptical. But the pointlike charm quark was heavier than a proton—something that had not occurred before, and was shocking to physicists, because no one expected a structureless quark to be heavier than a proton, which is made of other quarks. Within a year additional hadrons containing charm quarks were discovered, and soon after the Standard Theory was firmly established as the basic theory for particle physics.

### DISCOVERY OF GLUONS

One of the rules for calculating the production of quarks in electron-positron collisions is that quark and antiquark must always

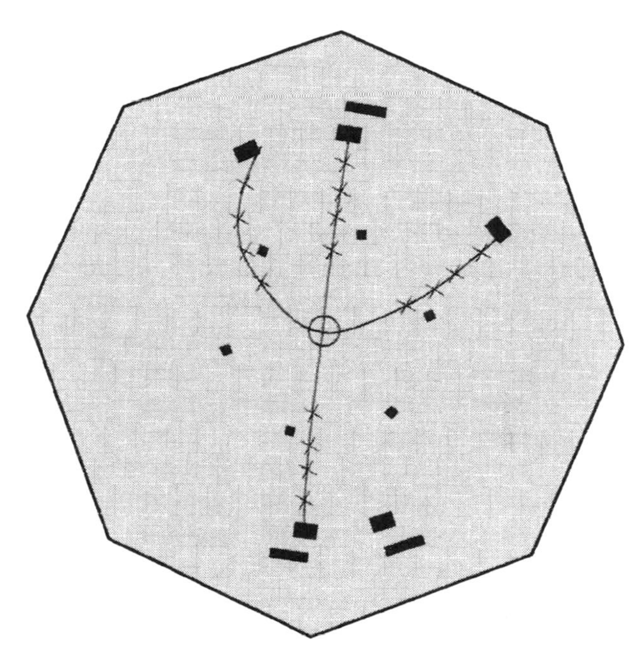

**Figure 6.1** *As we have seen, many of the particles discovered in the past have been given names represented by Greek letters. The discovery of the charm quark occurred through the discovery of new hadrons, simultaneously at SLAC and Brookhaven. After computer reconstruction, one early event showing a hadron decaying in the detector at SLAC looked as shown in the figure. The crosses and rectangles show where an electrically charged particle passed through a detector element. (The two curved tracks can be identified as pions, and the straight tracks as an electron and a positron.) The SLAC group (the MARK I collaboration) had proposed the hadron made of charmed quarks should be named ψ (Greek letter psi) and took this event as a sign that their name should be accepted. The Brookhaven group had proposed the symbol J. So this hadron is called "J/ψ," pronounced "J-psi."*

be produced together, so every event should show two jets. Since gluons carry color charge, they also have to appear as a jet of hadrons (remember, only color-neutral particles can emerge from collisions). It was soon realized that direct evidence for gluons should come from seeing events with a third jet, since

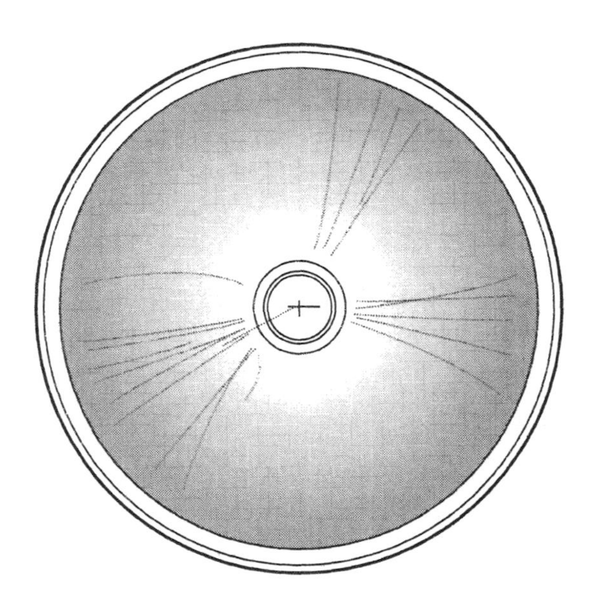

**Figure 6.2.** *This figure, originally taken by the JADE detector collaboratio at the PETRA collider at the German laboratory DESY in Hamburg, shov three jets of hadrons. Two are quark and antiquark, and the third is evidenc for the existence of gluons. In any particular event it is not possible to identi unambiguously which is a gluon, but with a large sample of events the ident fication can be made statistically.*

gluons (like photons) could be produced singly. Such events wer searched for, and first seen (Fig. 6.2) at PETRA in 1979, by sev eral detector groups. Both the rate and the direction of the gluo jets were in agreement with the Standard Theory predictions.

## TWO NEUTRINOS, AND THEN THREE

At Brookhaven in the 1960s a number of experiments wer done whose results were incorporated into the structure of th Standard Theory. One particularly important experiment led t the discovery that the muon had its own neutrino, differen from the electron neutrino. There was no clear reason to expec a second neutrino. But experimenters had searched for a particular muon decay, into an electron plus a photon, and not found it, even though it could occur according to what was known at the time. Theorists tried to understand that result, and suggested that a possible explanation could be that muons and electrons had some property that distinguished them besides their mass. In that case the neutrinos might also have such a property. If both neutrinos existed and were different, then when electron neutrinos hit a target, electrons should be produced but not muons, and when muon neutrinos hit a target muons should be produced but not electrons. A clever arrangement planned by Leon Lederman, Melvin Schwartz, and Jack Steinberger made this test possible. They and their collaborators did the experiment and found that indeed the electron and muon neutrinos were different particles. After the tau lepton was found a decade later, careful analysis of its decays allowed the deduction that the tau neutrino was also different from the electron and muon neutrinos. If those neutrinos had not been different, the symmetry structure of the theory would not have worked (see Appendix B).

### DISCOVERY OF THE W'S AND Z'S

By the middle 1970s, a number of leading particle physicists were convinced that the Standard Theory was a good description of nature. At that time the gluons, and the $W^+$, $Z^0$, and $W^-$ had been predicted but not detected. People expected the gluons to be discovered at the DESY collider, as they eventually were, but the electroweak gauge bosons (W's and Z's) were a bigger challenge. The problem was that the electroweak bosons were expected to be heavy, so heavy that it was hard to accept the prediction, and second, that no existing facility could hope to have enough energy to produce them. They were predicted to weigh about the same as a krypton nucleus. The krypton nucleus is a large composite object, composed of thirty-six

protons and about forty-seven neutrons, each of which is mad
of quarks and gluons. Yet the W and Z, of comparable weigh
are supposed to be completely pointlike, without structure.

Both Fermilab in the United States, and CERN in Switzerlanc
had proton accelerators that could be upgraded to produce W'
and Z's—if the theory were in fact correctly predicting thei
masses and production rates. If the theory were wrong an
W's and Z's had structure like a nucleus or even like a protor
their production rates would be much smaller, and any propose
collider would see no events. The United States showed no inte
est in searching for W's and Z's, but Europe did, initiating a serie
of decisions that has kept Europe at the forefront of particle phys
ics. CERN set out to construct a collider and detectors to fin
them. About eight years later (1983) a few W's and Z's wer
detected, as described near the end of Chapter 4. Every propert
was as predicted by the Standard Theory.

## MANY DIFFERENT EXPERIMENTS FIND THE SAME ELECTROWEAK UNIFICATION ANGLE

As described in Chapter 4, the electroweak unification angle $\theta$
enters into the description of many different experiments. Witr
out the Standard Theory, the quantity called $\theta_W$ in one exper
ment could have had a different value from the quantity calle
$\theta_W$ in another experiment. In fact, fifteen separate experiment
all found its value to be the same. That is an extremely powerfu
verification of the Standard Theory. It is a particularly crucia
one because there could have been no way to rescue the theor
if very different values of $\theta_W$ had been found in differer
measurements.

## MANY STANDARD THEORY TESTS AT LEP

In 1981 CERN began construction of an electron-positron coll
der (LEP) that would be able to produce millions of $Z^0$ bosons. I
is impressive that physicists and science ministers of CERN'

member countries had enough vision, and confidence in the Standard Theory, to initiate this project and begin construction before the existence of the $Z^0$ was confirmed in 1984. LEP began taking data in 1989 and has collected several million events that have been analyzed in detail. Study of the Z's and their decays made it possible to test the predictions of the Standard Theory tens of times more accurately than before. Many predictions have been confirmed to new levels of accuracy, including several that test whether the electroweak theory obeys the rules of quantum theory, and whether it is indeed a renormalizable theory (see the end of Chapter 4). Collectively, the results from LEP show that the Standard Theory describes nature very well.

### THE TOP QUARK

A major test of the Standard Theory was under way in 1994. The structure of the theory, combined with measurements of the b quark properties, has been known since 1982 to imply that the bottom quark must be part of a doublet, just as the up and down quarks, and the charm and strange quarks are—see Fig. 4.1. The top quark must accompany the bottom in the doublet. But the top quark had not been directly detected experimentally, and the Standard Theory is unable to predict the mass of the top quark. Measurements at Fermilab since 1992 have set a lower limit on the top quark mass, below which it would already have been detected. Measurements of other quantities at LEP, combined with the consistency requirements of the theory, have indirectly set an upper limit on the top quark mass. If the mass really lies in the allowed region implied by the LEP data, many top quarks will be detected at Fermilab (assuming the collider runs with expected intensity). Indeed, evidence for the top quark was reported in early 1994 by the CDF group at FNAL, and with more data the discovery of the top quark should definitely be achieved. In the spring of 1994, the CDF detector group published "Evidence for the Top Quark . . . ." The additional data taken in late 1994 and 1995 should lead to "Discovery of . . . ."

## NO EXPERIMENT DISAGREES WITH STANDARD THEORY PREDICTIONS

All the preceding experimental tests confirmed the Standar Theory, or participated in establishing it. A result that is perhap as important as any other is that no measurement disagrees wit the Standard Theory predictions, though many could have. Ther are two possible kinds of disagreements. One is the occurrenc of phenomena not predicted by the Standard Theory, such a a new particle (see Chapter 10 for examples) or the decay of proton (see Chapter 5). Rather than a failure of the Standar Theory, these would be welcome clues telling us how to exten the Standard Theory. The other would be a contradiction of th internal structure of the theory. For example, if different experi ments had given markedly different values of $\theta_W$, or if quark jet had not behaved as spin 1/2 particles, or if different $Z^0$ decay had occurred in the wrong proportion, or if dozens of othe things had happened, then the Standard Theory would just b wrong. None of the latter has happened. If the top quark wer not detected at Fermilab, the Standard Theory would have bee wrong.

# 7
# What Do Physicists Mean by "We Understand"?

Since we know in principle how to calculate the behavior of atoms, we can say we understand atoms. But if we have to just take the value of the electron mass as measured, rather than derive it, and since the value of the electron mass affects the properties of atoms, there is a sense in which we do not fully understand atoms. If we analyze what "understand" means, we can clarify how to describe how much understanding we have.

## *Levels of Understanding*

In order to explain what has been accomplished in particle physics, and the aims of current research, I have found it useful to distinguish three levels of understanding. The first and most familiar level can be called Descriptive Understanding. Two further levels, which I call Input and Mechanism Understanding and Why Understanding, help considerably to clarify the meaning of "understanding." Both are rather subtle, somewhat special to particle physics, and both are hard to find analogies for. The criteria for Input and Mechanism Understanding are especially unfamiliar ones.

Perhaps an imprecise analogy would be useful to illustrat the distinctions among the levels of understanding. If the worl is a videocassette recorder (VCR), and we can work the VCR we and do everything with it that it is capable of, then we have Descriptive Understanding. Of course no documentation cam with it—experimentation was essential to learn. But at this leve we don't understand it well enough to fix it if it is broken. Inpu and Mechanism Understanding would mean we could repair without outside help or parts. Why Understanding would mea we could invent the idea of a VCR and design it and make it fron raw materials without outside help. Clearly, Why Understanc ing is really vastly more difficult than Input and Mechanisn Understanding.

## *Level I—Descriptive Understanding*

A subfield or area of science (for example, particle physic atomic physics—the study of atoms, microbiology) has at tained a Descriptive Understanding of its subject matter it is possible to give a complete and well-tested descriptio of how things work in that subfield. It is necessary to b able to correctly predict the results of experiments, or a least be able to interpret experiments after they are done i the measurements involve too much complexity to carry ou a calculation ahead of time. For example, Newton provided Descriptive Understanding of motion and of gravitation, becaus with his laws it became possible to take various inputs such a masses and then describe the motion of all objects. With th Standard Theory, particle physics has achieved a Descriptiv Understanding of its entire domain, as have some other subfield of physics.

Descriptive Understanding in particle physics requires knowl edge of three kinds that we first met in Chapter 2 from th historical point of view, and then in Chapter 4 as a way o organizing the Standard Theory. First, to understand what we are made of, we must know what *constituents* are found when we subdivide matter into smaller and smaller pieces, and we must have good arguments that even smaller subdivisions are not possible. Of course, we can never prove that smaller subdivisions will not be found if a future supercollider can probe to smaller distances. But we have seen that matter has already been probed to distances far smaller than where structure might have occurred, so perhaps we do know the final constituents.

Second, we must know the *forces* that act on the constituents to bind them into the structures we encounter—quarks into protons and neutrons, protons and neutrons into nuclei, nuclei and electrons into atoms, atoms into molecules, molecules into us.

Finally, we must know the *rules* to use in order to calculate how the constituents interact under the influence of the forces, as described in Chapter 4. Einstein's special relativity and the quantum theory are the modern form of the rules. Since the 1920s the rules for calculating have not changed in principle, though there has been tremendous improvement in understanding how to use them.

The progress of recent decades has been in learning what the constituents and the forces are. Before the 1960s the electron was known, but there was no idea what the other basic constituents could be. Before the 1970s the electromagnetic and gravitational forces were known, including the proper expressions to use for them in Newton's law or the Schrödinger equation of quantum theory. The existence of the weak force was also known, but there was no understanding of what to put into the rules to calculate its effects. The existence of the strong force was unknown, though hinted at by the existence of one of its manifestations, the nuclear force. Today the situation is totally different—the Standard Theory satisfies all the conditions for a full Descriptive Understanding.

## *Level II—Input and Mechanism Understanding*

The second level, Input and Mechanism Understanding, is the least familiar one. It arises in the sciences, where the Descriptive Understanding is mathematical. To achieve this level it is necessary to recognize the mechanisms for how things work, which can include knowing how symmetries of the theory are broken (an example is given later in this chapter), or knowing which solution of an equation describes a system (see examples below). Input and Mechanism Understanding also requires that no masses, constants, or parameters be inputs taken from measurement or other subfields; they must be derivable from the theory of the subfield itself. Inputs (such as masses or force strengths) are not the same as mechanisms, so one could speak of an Input Understanding or a Mechanism Understanding, but their logical status seems to be the same, and sometimes they are connected (as with the Higgs mechanism and masses explained in Chapter 8), so probably they should be thought of as the same level of understanding.

For example, consider Newton's law of gravitation, which says that any two bodies are attracted to each other in proportion to the product of their masses. Many people asked what was really happening to cause the attraction—what was the mechanism? Newton did not know, nor did anyone else for over two hundred years. A Mechanism Understanding was not achieved until after the notion of a field was elucidated by Faraday and Maxwell in the nineteenth century, and finally Einstein showed that changes in the field (because of particle motion) travel at the speed of light. Then the "action-at-a-distance" problem was resolved, and the mechanism by which gravity worked was understood: A field is established by every massive body, and when it moves the field configuration propagates out from the body throughout space at the speed of light; any other massive body feels the field and is attracted.

When I was taught atomic physics and quantum theory, the

approach was that we were given electrons and nuclei with certain masses and other properties, and a force acting between them with a strength characterized by the size of the electric charge on each. The problem was to calculate the properties of atoms, such as their energy levels, various experimentally observable rates, the colors of light they emitted, and so on. All results were to be expressed in terms of a unit of mass such as the proton mass, a unit of electric charge such as the magnitude of the electric charge of the electron, a unit of time, and a unit of distance. The effort to do that for atoms had been successful, so we were taught that atoms were understood. Questions were not raised about *why* the force that bound the atom was what it was, *why* the proton and electron had the same magnitude of electric charge, *why* the masses were what they were, etc. It is still that way today; in each subfield of physics some quantities are imported and not explained.

Particle physicists have achieved a Descriptive Understanding of our world, but not yet an Input and Mechanism Understanding because it is still necessary to input particle masses obtained through measurements in order to predict the results of experiments. It is hoped (see Chapters 9 and 11) that eventually the values of the masses can be calculated as a part of the theory. The physics underlying the Higgs mechanism (Chapter 8) also has to be understood before an Input and Mechanism Understanding of particle physics is achieved. Of course, different parts of an Input and Mechanism Understanding can be achieved at different times.

As the Standard Theory was developed it turned out that the mathematical form of each force could be derived from the rules once it was guessed that the force exists (see Chapter 4), and even the *existence* of a new kind of force is required if a new type of charge is discovered. This was more intellectually satisfying than having the forces be completely independent of the particle properties and the rules, and it stimulated a desire to derive more aspects of the theory that had previously seemed independent.

Unfortunately, there has been little progress in deriving the values of the masses.

There are twelve quark and lepton masses, and two new kinds of charge (color charge and weak charge, described in Chapter 4) in addition to electric charge. Although the masses can be measured, their values cannot yet be predicted by the theory. Once the masses are measured, everything depending on them can be calculated. There are even some additional quantities (Chapter 4) that have to be measured before all phenomena in the everyday world and all experiments at accelerators can be incorporated into what is describable by the Standard Theory. (A caveat is needed here about things that are too complicated to calculate in practice—we'll wait till Chapter 13 for that.) Today in particle physics this lack of knowledge of these nearly twenty quantities is considered intellectually unacceptable, and a great deal of research effort is put into trying to extend the theory so that all masses can be calculated as ratios to one basic mass. That goal is often expressed by saying that all mass ratios, such as the ratio of the mass of the electron to the mass of the muon, and the mass of the electron to the mass of the top quark, should be calculable. Similar goals hold for the other parameters, such as interaction strengths. Achieving those goals would take us beyond a Descriptive Understanding.

It takes at least three fundamental constants of nature to express everything else as ratios; they fix the units of measurement. They can be chosen in various ways, but the favored choices today are usually to use Newton's gravitational constant, **G**, Planck's quantum unit, **h**, and the speed of light (which we could call Einstein's constant), **c**. Although we will not use them much in this book, it is worth explaining them a little since they are very basic. Input and Mechanism Understanding essentially implies that all other measurable quantities can be expressed in terms of these three constants, or equivalent ones.

Newton's law of gravitation says that any two objects feel a mutual gravitational force that is proportional to the product of

their masses divided by the square of the distance between them. In order to complete the specification it is necessary to say how strong the gravitational force is for two particular masses a specified distance apart, which requires multiplying the other factors by a constant (which we call **G**). That turns the proportionality into an equation. Similarly, one way to characterize the basic innovation of the quantum theory is to say that the energy levels of atoms are quantized rather than continuous. That is, they are only allowed to take certain values and not those in between. The spacing between levels is determined by Planck's constant, **h**.

Einstein's theory of special relativity follows from two assumptions. The first is that the formulations of the basic laws of nature should not depend on whether they are studied in a laboratory located here on earth, or in an airplane, or somewhere in outer space, or in any other frame moving at some speed relative to these. No one should argue with that. The second is more remarkable. It is that the speed of light (in free space) has the same value under all conditions. In particular, it does not depend on the speed of the source that emits the light. That is very different from our everyday experience. If we observe a person throwing a ball with some speed, and then observe her throw it again while passing by us in a car, we know that in the second case the speed of the ball relative to us is the combined speed the throw gave the ball plus the speed of the car; it is not the same speed in both cases. Light behaves differently (as do other things moving with nearly the speed of light). All light travels at a speed **c** (c is about $3 \times 10^8$ meters/second), whether it is emitted by a stationary flashlight or by one in motion.

## MECHANISM UNDERSTANDING—BREAKING THE SYMMETRY OF THE EQUATIONS

There is a second aspect to Input and Mechanism Understanding for particle physics. Although we really want to formulate a set of equations that constitutes the underlying theory, what we see around us and in experiments is described by the solutions to the equations. The equations (hopefully) have considerable simplicity and symmetry, but the solutions usually do not. Often equations have many solutions; before it is possible to make predictions it is necessary to know which solution corresponds to our universe.

A simple (artificial) example may clarify this situation. Suppose the basic equation from the theory that predicts the masses of the electron, muon, and tau leptons is $E \times M \times T = 64$, where E stands for the mass of the electron, M for the mass of the muon, T for the mass of the tau, and the equation just says multiply the three masses and the product is 64 (in appropriate units). This equation has a number of solutions. For example, all the masses could be equal, E = M = T = 4. Or two could be equal and lighter, E = M = 2, T = 16. Or two could be equal and heavier, E = 1, M = T = 8. And which two are taken equal could be different, e.g. E = 16, M = T = 2. They could all be different, E = 2, M = 4, T = 8, or E = 1, M = 4, T = 16. In physics, of course, the equations are much harder to solve. One problem is to solve them at all, to find out if any solution looks like our universe. One might work hard for months to get a solution that turns out not to look like our universe—although maybe a different solution would. Even if it did look like our universe it would be necessary to show that it was indeed the right solution—why that one instead of a different one?

Note the important feature that the original equation for the masses is completely symmetric. It looks the same if you interchange the symbol for the electron mass with either of the other masses, or the symbols for the muon and tau masses, etc. The equation does not distinguish between the electron and the muon and the tau. But the solutions can give them very different masses. Today we suspect nature is like that—the Standard Theory equations are symmetric in many ways (some are described in Appendix B). The equations have a number of different solutions. The world is described by a solution that is not symmetric. If that could be derived, it would take us to an Input and Mechanism Understanding; some work in that direction is the subject of the next chapter, Higgs physics.

The jargon physicists use to denote describing the world by a nonsymmetric solution of a symmetric equation is "spontaneous symmetry breaking," and the state arrived at by doing so is called the "vacuum" selected by the spontaneous symmetry breaking. Input and Mechanism Understanding requires explaining why the vacuum state of our world is what it is in addition to knowing the equations of the theory. This is a new consideration in the history of physics. It occurs in the Standard Theory and in all attempts to extend the Standard Theory, and in other areas of physics, such as condensed matter physics.

## Level III—Why Understanding

The third level of understanding—the"Why"—is the most ambitious. Having a Why Understanding would require being able to explain *why* things are the way they are, and *what* things really *are.* For example, claiming that a Why Understanding of particle physics was achieved would require a successful derivation of the existence of the electron, muon, and tau leptons, the values of their masses, and an explanation of why three (and only three) such particles that differ only in mass should exist. It would require explaining *what* electric charge is, *why* the forces are the ones we know and not others, and much more. While a Why Understanding is a fine goal, it may never be attained. As a proverb says, "The mere existence of a problem is no proof of the existence of a solution." However, achieving a Why Understanding of particle physics is a major research activity today (see Chapter 11), an attempt at a complete theory of "everything." The goal of achieving such a theory is ancient, but it has only become a practical research topic since the middle 1980s.

Historically, Why Understandings have been the domain of religions and mythologies. Believers often requested favors of the principals of their creation myths, although they never requested information as to the creators' construction techniques. They couldn't because they didn't have enough knowledge to understand the questions, let alone the answers. Having gained a Descriptive Understanding—a knowledge of how the world works—we are asking now for knowledge of the tools and the pattern. As we will see in Chapter 11, whatever the eventual outcome, the search for a Why Understanding is now legitimately in the domain of science.

Some people have argued that a full "theory of everything" can never be achieved by humans. And some people feel the human mind cannot comprehend the universe it is a part of—though no one has ever given a scientific argument for that belief. Some people think that throughout history every scientific development opened up new questions and that process will continue indefinitely, so we will never reach final answers. Again, there is no convincing argument for that belief, and perhaps the analogy of the exploration of the surface of the earth is relevant—that geographic exploration continued for millennia, with every foray leading to further vistas, but it finally ended in the twentieth century. And some people are uneasy about religious implications of a theory of everything.

In practice, of course, the quest will go on. Since Galileo science has developed one step at a time. Those who consider understanding the universe important will push as far as they can. The glimpses so far of how nature works have shown an austere beauty, hinting at more. Perhaps a limit will be reached because of intellectual or technological or economic limitations, or because society considers the expense of experiments too great for the value given to doing science and understanding nature. Or perhaps eventually we will get there.

# 8
# Higgs Physics

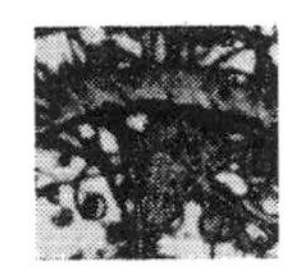

Higgs bosons (if they exist) are the quanta that mediate a new kind of interaction, one that gives rise to mass. By about 1915 electrons and photons were known to exist. The electron was the first of the twelve fundamental fermions (leptons and quarks) to be discovered. The photon was the first of the twelve bosons (photon, $W^+$, $Z^0$, $W^-$, eight gluons) that transmit the forces of the Standard Theory to be discovered. The Higgs boson predicted to exist by the Standard Theory is a new kind of particle different from either of those.

In fact, a manifestation of the physics that leads to Higgs bosons has already been observed, but subtle properties of the Standard Theory have conspired to prevent us from interpreting it. Light (photons) can be polarized in two directions, a phenomenon familiar from sunglasses that use materials that do not transmit one polarization direction in order to reduce the intensity. A similar property holds for gluons. W's and Z, however, are massive—they get their mass by the "Higgs mechanism," described below. Massive bosons have three polarization states instead of two. The third state is incorporated via the Higgs physics. Since W's are studied at Fermilab (and after 1996 at LEP), and Z's at LEP, we could hope to study the third polarization states, which are truly a new kind of particle. But, sadly, the requirements of special relativity and of quantum theory fix the properties of the third polarization state so completely that we cannot learn about the underlying Higgs physics by studying a single W or Z. Studying the collisions of two W's or two Z's at a sufficiently large energy is required to make progress, and the possibility of doing that experimentally is not guaranteed at any planned facility.

## *Why?*

The motivation for the prediction of the Higgs boson is not the normal motivation for predictions in much of physics. If Higgs bosons do not exist, there is no disagreement of a theoretical prediction with an existing experiment, no empirical puzzle. What is at stake is instead the existence of a meaningful theory. In its basic form, the Standard Theory is a theory for massless particles. All the leptons, quarks, and bosons must be particles without mass, or the mathematical consistency of the theory is destroyed. The photon and the gluons indeed have no mass, but all the others do. Why not just insert a mass for them in the equations? Unfortunately, in a quantum theory all aspects of the physics are so highly interconnected that if the masses are just put in, then calculations start to give infinite values for the predictions for many ordinary measurements. (In the language of the last section of Chapter 4, the theory is then not renormalizable.)

But during the 1960s it was learned that the theory of massless particles could be modified by adding to it a new kind of field, the Higgs field. Then the underlying structure of the theory was not changed from that of the massless theory, and it remained renormalizable; thus all experimental quantities could be calculated. According to the theory the masses of particles arise from their interactions with the Higgs field, and thus all masses can be included in a consistent manner. The process by which the particles get mass is called the "Higgs mechanism."

Just as the quanta of the electromagnetic field are the photons (the bosons that mediate the electromagnetic interaction), the Higgs field has quanta—the Higgs bosons. Although the Higgs mechanism allows masses to be included in the theory, it is not powerful enough to predict the actual values of any quark or lepton masses, or the Higgs boson mass itself, but it is able to predict (correctly) the masses of the W's and Z's in terms of other measured quantities—and thus numerically. (The prediction of the W and Z masses is a major success, but it is not deeper than a Descriptive Understanding because those masses are only related to other quantities whose measured values are used.)

The Higgs field has one particularly unexpected property. In physics the state in which all fields have their lowest energy is called the vacuum. For most fields—for example, the electromagnetic field—the energy is minimized when the value of the field is zero everywhere, so the physics vacuum has nothing permanent in it; it is the empty vacuum that would naively be expected. For the Higgs field, theorists conjectured that the energy associated with the field is actually smaller when the field has a constant value (different from zero) everywhere than when it just vanishes. Thus the Higgs field does not vanish in the vacuum. This result and its effects are called the Higgs mechanism. (The jargon for preventing the field from vanishing in the vacuum is called giving the Higgs field a "vacuum expectation value.")

The Standard Theory with a Higgs field is a completely consistent relativistic quantum field theory and is totally satisfactory by all historical standards in physics. The solution it provides to including mass in the theory is technically satisfactory and is superior to most proposed alternatives. It successfully introduces mass for W's and Z's and the quarks and leptons. Nevertheless, many particle physicists are uneasy about it. Some are uneasy because the whole approach has an ad hoc nature—the Higgs field is introduced to do its job and does it, but several assumptions are needed, and what is added has few observable implications that allow other tests of its presence. Others are uneasy because of subtle theoretical implications of the Higgs physics (these will enter in Chapters 9 and 11). In general there is uneasiness because no one understands what gives rise to a Higgs field. The electromagnetic field arises from electric charges, and the color field from color charges, but what gives rise to a Higgs field?

In this chapter and some subsequent ones a new feature will enter compared to the presentation so far. Up to this stage the Standard Theory is a tested, established body of experiment and theory. From here on we set off through less certain territory. Alternatives and opinions will enter because this is research in progress.

## *Discovering a Higgs Boson*

The question would be settled, of course, if Higgs bosons were discovered, or if experiment demonstrated that they did not exist. The problem is that because the theory does not predict the mass of the Higgs boson, experiments must search for a Higgs boson of all possible masses until it is discovered or shown not to exist. Until the LEP collider at CERN began to take data in 1989, no experiment that was likely to detect a Higgs boson had been possible. Now we know that if a Higgs boson had a mass lighter than about two-thirds of the $Z^0$ boson mass, it would have been already detected at LEP; if it were heavier than that it could not yet have been found.

How heavy should a Higgs boson be? In the Standard Theory, one can show that if a Higgs boson exists then its mass must be less than about eight times the Z mass. With a certain plausible assumption this region shrinks considerably and the Higgs boson mass must be below about two times the Z mass. Unfortunately, the LEP collider cannot search up to these values. If a Higgs boson is detected at LEP the question is settled, but if one is not detected there we don't learn new definitive information.

Both the CERN LHC, and the NLC that several countries are

considering building (Japan, Germany, the United States), could eventually cover this full mass range and either detect a Higgs boson if it exists, or demonstrate that it is not there. The FNAL collider could be upgraded in luminosity and detect or exclude a Higgs boson in the most interesting range of masses.

## *Attitudes About Higgs Bosons*

Almost everyone who has thought carefully about Higgs physics is dissatisfied with the way the Standard Theory formulates it. But it is possible to prove that if the simplest form of a Higgs field with Higgs boson quanta is not nature's way, then some other physics must occur that has the same effects for generating the masses of quarks and leptons and bosons. Because of that constraint, the question of Higgs physics cannot be ignored. Contenders then fall into three basic camps.

A few have said they think the idea of a Higgs field and an associated Higgs boson is just not right, and that some other unknown mechanism will replace it. Proponents of this approach (we can think of them as atheists) have argued that when the interaction of a pair of W bosons is someday studied they expect new effects to be observed. Unfortunately, even the most energetic collider planned, the CERN LHC, will not have sufficient energy to perform a general study of this alternative.

Another small group (agnostics) have said they also do not believe there is a fundamental Higgs field and a Higgs boson. They think that instead of Higgs fields several new matter particles (quarks and leptons and perhaps additional types) will be found to exist, and that the new particles will be found to carry one or more additional charges (like color charge but with more allowed values than three). Then new bosons and new interactions must also exist. The new interactions lead to new types of hadrons made of the new particles, and some of the new hadrons can play the role of Higgs bosons. A variation has the new interactions leading to a particularly strong interaction of top quarks (which is special because the top quark is the heaviest of all the Standard Theory quarks, leptons, and bosons); a hadron composed of top and antitop then behaves like the Higgs boson. These are clever and interesting ideas, and they still have strong proponents. However, they have been less successful than hoped, particularly for two reasons. One is that they are as effective as the normal Higgs mechanism in introducing mass for W's and Z's into the theory, but they fail (or require very contrived assumptions) to deal with the masses of the quarks and leptons. Second, they have to claim that some major successes of alternative approaches (described in the next two chapters) are accidental, and without scientific significance, since their approach lacks those successes.

The majority of people (fundamentalists) who have thought about Higgs physics believe that fundamental Higgs bosons really do exist, but that the Standard Theory has to be embedded in an extended theory in which the Higgs physics is less arbitrary than in the Standard Theory. Fortunately such a theory already exists, the "supersymmetric" Standard Theory. Some proponents of this approach come to it because of several attractive features of the supersymmetric theories themselves, others because the Higgs physics is part of the supersymmetric theory (see Chapter 10), and even more because "theories of everything," such as superstring theories, that seek to explain our world automatically contain the supersymmetric Standard Theory. In the next two chapters we will survey the motivations for extending the Standard Theory even though there are no experimental indications that it needs extending, and then describe the gains achieved by adding supersymmetry to the Standard Theory. Viewing the Standard Theory from the supersymmetry perspective dramatically changes the approach to Higgs physics.

In a supersymmetric theory it is still necessary to assume that Higgs fields exist, but once that is accepted most of the arbitrariness of the Standard Theory is gone. Most important, in the Standard Theory extra assumptions are needed to activate the Higgs mechanism that gives masses to the other particles, but in the supersymmetric theory the Higgs mechanism emerges as a derivable result in an elegant way. Further, while the mass of the Higgs boson is not calculable in the supersymmetric theory either, its allowed range is more constrained and in particular has a firm upper bound that is well below even the twice the Z mass bound of the Standard Theory. The upper bound on the supersymmetric Higgs boson mass is at the technical limit of what could be searched for at the CERN LEP collider, though expensive upgrades would be needed to extend the capabilities of the collider that far. However, the actual mass is not necessarily near the upper limit, so finding the Higgs boson at the CERN LEP collider is a good possibility if nature is described by a supersymmetric Standard Theory. The proposed CERN proton collider (LHC) would produce large numbers of the Higgs boson of the supersymmetric Standard Theory (if the latter exists and the former is eventually built). At that facility (for technical reasons) finding the Higgs boson is possible but difficult, requiring special detector capabilities; whether appropriate detectors will be built is not yet known. If the FNAL collider were upgraded sufficiently in luminosity, it would be able to detect the supersymmetric Higgs boson.

The Standard Theory will only be complete when the Higgs physics aspects are settled. Finding out whether the "atheists" or the "agnostics" or the "fundamentalists" are right requires experiments that either do or do not discover Higgs bosons, and that carefully study the properties of any new particle that might be a Higgs boson. No amount of arguing will decide (though there has been and will be a lot of arguing)—only data. At the same time, as the alternatives described hint, the outcome of the effort to understand the Higgs physics will point the way to extending the Standard Theory.

# 9
# The Standard Theory Will Be Extended

If the Standard Theory is as good as the previous chapters say, why don't particle physicists close up shop and work on something harder, or write books? There is, of course, a great deal to do to understand the properties of the solutions to the basic equations, and all of the implications and predictions of the theory; many particle physicists are working in these directions. But there are also very strong arguments that the theory will be extended. To be more precise, there are four very strong arguments based on observed phenomena and another three arguments based on hope. The first two strong arguments are closely related but independent, the third is somewhat related to them, while the fourth is very different.

## *Phenomena Not Predicted by the Standard Theory*

### A NEW SCALE IN NATURE

There is a remarkable implication of the Standard Theory, first noticed in the mid-1970s by Howard Georgi, Helen Quinn, and Steven Weinberg, that seems to imply that the Standard Theory will be embedded in a more encompassing theory. Because the Standard Theory is a mathematical theory, we can ask what the weak, electromagnetic, and strong forces would be like if we could do experiments at smaller and smaller distances; colliders are like microscopes, with higher energy ones probing to smaller distances. Even though we do not have the experimental facilities to do that, we can calculate the predictions of the Standard Theory for such a question. The result is that at shorter distances the three forces look more and more alike, and at a distance of about $10^{-30}$ cm the forces seem to become the same, to "unify." Qualitatively, at least, the three forces besides gravity have different effects at larger distances, but perhaps are just different manifestations of one underlying force, much as electricity and magnetism were seen, over a century ago, to be different manifestations of one force.

This result has powerful implications. It need not have happened, in the sense that nothing we know about the Standard Theory itself requires this coming together of forces. It depends on two things: The first is the way the Standard Theory tells us to extrapolate the behavior of the force strengths to smaller distances; the second is the starting values for the extrapolation for each force strength. The latter are measured in experiments, and could have been very different without contradicting the structure of the Standard Theory in any way we know. For example, from the point of view of the Standard Theory the strength of the electroweak force is described by a quantity called $\alpha_2$. In the Standard Theory $\alpha_2$ is simply something to measure. It comes out to be $\alpha_2 = 0.033 \pm .003$. If it had turned out to be half that value, or twice it, the unification would simply not occur.

Of course, someone could claim that the coming together of the forces was a meaningless coincidence, and that is possible, though unlikely. Also, detailed study produces further support for the significance of this result (as described in the next chapter). What it is telling us is that there is a new scale in nature for which there was no previous evidence. As far as we know, the Standard

Theory would be entirely consistent and meaningful without this new scale. The existence of this new scale implies the laws of nature are even simpler at $10^{-30}$ cm than at the distances we know about directly from experiments (down to about $10^{-16}$ cm).

There is another implication. The equations of the Standard Theory are much too complicated to be solved explicitly by any technique we know today. In order to do the extrapolation to shorter distances a method is used that gives an approximate solution. (The jargon used to describe that method is "perturbative.") It could have happened that nature is such that the assumptions used to justify the approximate solution are wrong. When that happens in other areas, usually the outcome is complicated and hard to interpret. But here the outcome is a very simple, elegant result: The different forces become equal. That suggests that the approximations are valid, and that between the distances probed in experiments at LEP and Fermilab (about $10^{-16}$ cm) and the distance where the forces seem to merge (about $10^{-30}$ cm) there is no intermediate scale where major new phenomena occur—if there were one, the approximations should not have worked. This argument is not a proof, it is an indication. Whenever one is doing research in an unfinished area one has to look at clues and decide whether to take them seriously.

The extrapolation from $10^{-16}$ cm to $10^{-30}$ cm can seem a long one. But that depends on how it is viewed. Figure 9.1 shows the distance scales for a number of things we know about—that is, the sizes of the things, on a scale where each mark on the axis is ten times bigger or smaller than the adjacent one, a "logarithmic scale." Viewed this way, the distance from the weak interaction scale to the new scale, which is usually called the "grand unification" scale, or perhaps just the "unification" scale, is not much larger than other distances. This is not just a gimmick to make it look good—the Standard Theory tells us that a logarithmic scale is the right way to view it.

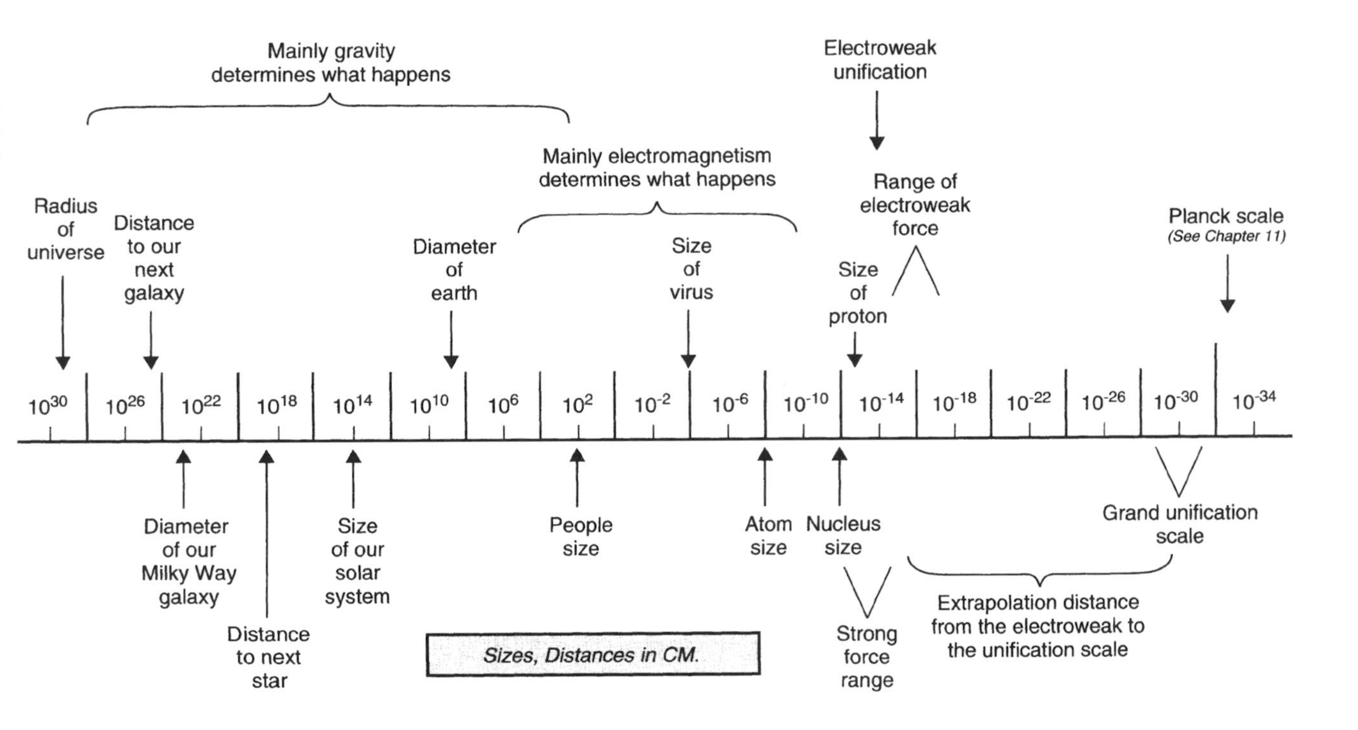

Mainly gravity determines what happens
Electroweak unification
Mainly electromagnetism determines what happens
Range of electroweak force
Radius of universe
Distance to our next galaxy
Diameter of earth
Size of virus
Size of proton
Planck scale
(See Chapter 11)
10$^{30}$
10$^{26}$
10$^{22}$
10$^{18}$
10$^{14}$
10$^{10}$
10$^{6}$
10$^{2}$
10$^{-2}$
10$^{-6}$
10$^{-10}$
10$^{-14}$
10$^{-18}$
10$^{-22}$
10$^{-26}$
10$^{-30}$
10$^{-34}$
Diameter of our Milky Way galaxy
Distance to next star
Size of our solar system
People size
Atom size
Nucleus size
Grand unification scale
Strong force range
Extrapolation distance from the electroweak to the unification scale
Sizes, Distances in CM.


FIGURE 9.1

## CALCULATING THE ELECTROWEAK UNIFICATION ANGLE $\theta_W$

Sometimes one can learn whether a clue is significant by pushing on it and seeing how it reacts. If one insists on taking the new scale seriously and doing the approximate calculation as well as possible today, then the results are even more interesting. Once more precise measurements from LEP became available in 1990, the extrapolation to short distances was repeated; it turns out to fail if one uses just the formulas of the Standard Theory! Earlier the data had been too imprecise to determine that. But a small change in the formulas that embed the Standard Theory in a particular extended theory (the "supersymmetric" one described in the next chapter) turns out to restore the unification. Other extended theories that work could be found too, so that does not imply the supersymmetric one is right, but it is strong encouragement to study it. In particular, starting from a supersymmetric, unified theory the angle $\theta_W$ described in Chapter 4 can now be predicted rather than just measured, and the prediction agrees with the measurement to an impressive accuracy of about 1 part in 200. All measurements have some uncertainty, and all theoretical predictions, which themselves depend on other measured quantities, also have some uncertainty. How well predictions can be tested is limited by the size of these uncertainties. After much work by both experimenters and theorists, the uncertainty in $\theta_W$ has been reduced to this small level. Such agreement seems likely to be a further clue that the Standard Theory will be extended—if not, this accurate agreement would also have to be a coincidence.

## QUARKS AND LEPTONS COME IN IDENTICAL PATTERNS

As we saw in Chapter 4, quarks and leptons seem to fall into identical patterns—the only difference was that quarks carried the color charge, and leptons did not. The Standard Theory does not require that they come in identical patterns. That they do again suggests that somehow they will be related in an extended theory, and that such a theory will be found.

### THE STANDARD THEORY HAS NO CANDIDATE FOR DARK MATTER

An amazing result, deduced from cosmological arguments and astrophysical data in several ways during the past two decades, is that most of the matter in the universe is not made of protons and neutrons and electrons at all, but something entirely different. Protons and neutrons and electrons condense under the action of the force of gravity and form stars, which clump into galaxies, which we can see. A number of features of the observed universe, particularly the motions of stars and galaxies, imply that it contains a certain amount of matter. But the amount visible in stars is only 1 or 2 percent of the total needed, and other arguments imply that less than 10 percent of the total could be the forms of matter that we know about. The rest of the matter in the universe is "dark," that is, it does not form stars that we can see.

This is a remarkable conclusion, and is not yet established beyond all doubt, but slowly over a decade loopholes and alternatives have been eliminated. If it is true, the next question is: What is the dark matter made of? That is even harder to establish than its existence. Increasingly, bits of evidence have been pointing toward the conclusion that the dark matter is made of as yet unidentified stable particles that exist in large numbers throughout the universe.

If the dark matter particles exist, we can expect a complete theory to include them along with the quarks and leptons and bosons. Although it describes our visible world, the Standard Theory has no room for dark matter particles. To many physicists that is another clue that the Standard Theory will be extended.

While there is no consensus about the result, it is possible to construct a theory with a dark matter candidate that does the needed job. Indeed, that can be done in the same supersymmetric extension of the Standard Theory that allows the calculation of $\theta_W$, described earlier. Even better, the dark matter candidate of the supersymmetric theory came naturally from the theory itself, rather than in response to the astrophysical observations of dark matter.

## *An Input and Mechanism Understanding of the Standard Theory?*

The Standard Theory is a description of how the laws of nature work. Once the masses of the quarks and leptons, and the strengths of the forces are measured, the values of these quantities can be used in the Standard Theory equations to calculate how the particles behave in experiments. Particle physicists hope (the first hope) that it is possible to do better. We want a theory that also predicts the values of the masses and force strengths. Discovering that theory would give us what was called an Input and Mechanism Understanding in Chapter 7.

Although almost all particle theorists believe that an Input and Mechanism Understanding will be achieved, no clear scientific argument exists to reassure us. Almost everyone agrees that the problem is not that the values of the masses could be calculated if we understood the existing theory better, but that a new principle which is part of an extended theory is needed. The role for experiment is not just to obtain the values of the masses, which are already measured except for neutrinos, but to provide data that points the way toward the new theory.

One part of attaining an Input and Mechanism Understanding is to learn the basic physics of the Higgs mechanism. Since the Higgs mechanism is in practice the way the masses are introduced into the equations, we hope (the second hope) that its physical basis will be understood, and when it is we will have a clue that suggests or selects the new theory that predicts the values of the masses. Progress here probably requires detecting and studying a Higgs boson, or demonstrating that none exists.

We can ask more generally what the ingredients of a healthy experimental program should be. If only scientific criteria are considered, not perceived funding considerations, and if we made a minimal list of the information about nature that would have the most impact on the development of particle physics, what would be on that list? Most clear is the need to discover a Higgs boson, or demonstrate that none exists; that is an opportunity for the CERN LEP collider and for Fermilab if it upgrades the collider luminosity, and then for the CERN LHC and the NLC. The superpartners described in the next chapter should be found or excluded—upgraded LEP and Fermilab colliders have a chance at that, and LHC and NLC will do well. If superpartners are found, the future direction of the field is settled; if superpartners do *not* exist, the majority of particle theorists are working in the wrong direction. It is very important to measure the top quark mass and properties well and determine whether the indirect determinations of the top mass agree with the actual mass. It is essential to learn about neutrino masses, and whether protons decay (and, if so, what they decay into). Discovery of any rare occurrences of decays forbidden by the Standard Theory would have a major impact on our thinking. A better understanding of CP violation (Appendix C) could be very important in determining how to proceed. Detecting dark matter particles in laboratory experiments, or demonstrating they do not exist, should be a major particle physics goal.

## A Why Understanding of the Laws of Nature?

Why are the universe and the laws that govern its behavior and evolution what they are? Why are there the four forces, and what is the connection of gravity to the weak, electromagnetic, and strong forces? Why are there three families of quarks and leptons and not fewer or more? Why are the rules specified by quantum theory and special relativity what they are?

Before we had the Standard Theory it would have been idle speculation to discuss such questions. Beginning in the middle 1980s, for the first time in history these kinds of questions, and more, such as the origins of space and time, have become research problems. In the past, once a topic became a research problem it was eventually solved, so perhaps that is the best bet (the third hope) for these questions too. Of course these are new kinds of questions, so history may not be a good guide. Sometimes people argue that humans will not be able to understand such questions, but no scientific argument has ever been given implying that; so far such arguments say more about the person arguing than about the topic.

A more serious question is whether the factor limiting our understanding, if there is one, will be a lack of experimental data to point the way. Without data to help decide among competing ideas, or to convince theorists to take a particular approach seriously and explore its implications, the connection to nature that provides the criterion for what is correct is gone.

Perhaps what is most likely is that the experiments will eventually be possible and will get done, but they will be hard and give limited information that has to be carefully interpreted. Good theories have a way of suggesting experiments or phenomena that could not have been imagined before the theory, such as the prediction of electromagnetic waves based on Maxwell's equations, confirmed twenty-five years later by Heinrich Hertz, or Einstein's explanation of the precession of the perihelion of Mercury, or the prediction of the existence of a certain kind of dark matter in supersymmetric theories (explained in Chapter 10); some more examples will be given in Chapter 11. If the arguments earlier in this chapter about a new scale in nature and the validity of the approximate calculations connecting quantities from that scale with things measured in today's experiments are correct, it means we can reliably extrapolate much of the way to the very small distances where today's ideas suggest the Why Understanding should apply. We can reasonably hope for experimentally detectable effects from the unification scale. Theorists around the world are presently involved in postulating models at the unification scale and extrapolating to predictions for experiments in progress; others are looking at existing data and trying to extrapolate to see what it implies for the unification scale.

# 10

# Supersymmetry—the Next Breakthrough?

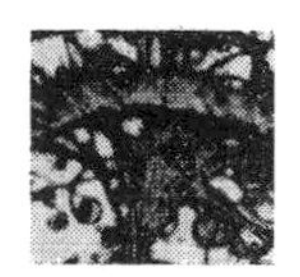

The previous two chapters introduced two strong clues that point to how to extend the Standard Theory—the Higgs mechanism works well technically to introduce mass into the theory, so it or something like it should be a consequence of an extended theory, and there is an apparent unification of forces at a distance of about $10^{-30}$ cm. These are the two things closest to "smoking guns" that we know of, though their interpretation is not yet clear.

## *What Do We Gain If the Theory Is Supersymmetric?*

The existence of a new and very different scale where physics is simpler puts very strong conditions on any theory that incorporates it. One can make very general (and unfortunately also very technical and complicated) arguments in a quantum theory that physics should not exist at two widely separated scales connected by the kinds of approximations described in the previous chapter, unless additional conditions are met. A separation of scales is natural only if the theory has certain properties called symmetries, such that various quantities combine just the right way to achieve the needed outcome.

A third possible clue as to how to extend the Standard Theory is the existence of the dark matter, which was briefly described in the previous chapter, and a fourth clue could be the suspiciously similar pattern of quarks and leptons. Further, the Standard Theory forces are not yet unified with gravity, and we would hope that the next extension of the Standard Theory at least has connections to gravity that might help us learn how to relate gravity to the Standard Theory forces.

There is another set of possible clues from experiment that can be called "The dog didn't bark" ones, after the essential clue used by Sherlock Holmes to solve the mystery of the disappearance of *Silver Blaze.* ("Obviously," Holmes says, "the midnight visitor was someone whom the dog knew well.") Similarly, many experiments have been carried out with the hope of finding a result showing the presence of physics not explained by the Standard Theory, only to end with a confirmation of the Standard Theory. Most extensions of the Standard Theory lead to a number of predictions for observations that are different from the Standard Theory predictions. The requirement that a satisfactory extension of the Standard Theory give nearly the same results as the Standard Theory for some measurements, but not for others, is a powerful constraint.

The reason many researchers are excited about the extension of the Standard Theory called the supersymmetric Standard Theory is that it satisfies all these requirements: it incorporates the Higgs physics; it accommodates the two widely separated scales in a consistent way; it can unify quarks and leptons and the Standard Theory forces; it has a dark matter candidate; it has a connection to gravity; and it has not "barked" when it shouldn't.

Nevertheless, until explicit experimental discoveries confirm the predictions more directly, we are far from knowing that nature is in fact supersymmetric. Later in this chapter I'll describe some of the discoveries that could occur. Perhaps a disclaimer is also appropriate: In case the reader has not guessed, I myself am an advocate of a supersymmetric theory, and most of my research is related to it. Not every physicist would give such a prominent position to supersymmetry today.

## *What Makes a Theory Supersymmetric?*

With the achievements already described in mind, let us see what makes a theory supersymmetric, and look at some of the implications of the supersymmetry. Near the end of Chapter 4 the property of particles called spin was described. All particles fall into one of two categories; they are fermions (if their spin is a half-integer multiple of a basic unit), or bosons (if it is an integer multiple); in practice, for particle physics, fermions have spin 1/2 and bosons have spin 0 or 1. It may seem a small difference, but it has profound implications for their behavior in a world where the rules are set by a quantum theory. One of many differences is that two bosons can occupy precisely the same place at the same time, while two fermions cannot. If everything were made of bosons, tables and walls would not be so solid. (These are fascinating effects, but would take us too far afield from our main topic, so I will not discuss them further.)

In the Standard Theory, matter particles (electrons and other leptons, and quarks) are fermions, while the quanta that transmit interactions are bosons. The basic idea of supersymmetry is to hypothesize that the fundamental law(s) of nature are really symmetric between bosons and fermions. Every Standard Theory boson has a fermion "superpartner" and vice versa. Superpartners are identical in every way (same mass, same electric charge, same color charge, etc.) except for their spins. The spins always differ by 1/2. The superpartner of the photon has spin 1/2, and the superpartner of the electron has spin 0.

With a mathematical theory one cannot just say, "Let's add a bunch of particles to the theory." It is necessary to extend the mathematical structure of the Standard Theory to incorporate all the superpartners and all the relations among them implied by

the new symmetry. It could easily have happened that an inconsistency appeared while constructing the larger theory. For supersymmetry, it worked—in the early 1970s several people were able to construct a supersymmetric relativistic quantum field theory.

Since the superpartners are not observed, while the ordinary particles are, the full supersymmetry must somehow be broken, even if it holds for the fundamental laws. If there were a partner of the electron, with the same mass as the electron, it would combine with nuclei to make atoms, and we would know about those atoms. If nature is supersymmetric, we must suppose that the superpartners are heavier than the particles we know, to account for why no existing facility has projectiles energetic enough to produce them so far. It could be like the oversimplified example of Chapter 7, where the equation to describe a particle mass (call it m) and its superpartner mass (call it M) is (in a simple unrealistic example) $mM = 16$. The equation is symmetric—it does not change if m and M are interchanged. The solution could be $m = M = 4$, or it could be $m = 2$ and $M = 8$. Apparently nature has chosen the latter, though we do not know why. Since mass is the least understood aspect of today's theories, we are not uncomfortable with having the supersymmetry broken only for the masses. All other aspects of the theory remain symmetric, so it is possible to calculate how the superpartners would interact, how many would be produced at a new collider, how they would decay, and so forth.

The supersymmetry theory tells us that additional numbers can be associated with all particles, in such a way that positive numbers are associated with ordinary particles, and negative numbers with superpartners. The theory also says that the product of all the numbers of the particles in a process cannot change sign. Therefore, if a process begins with ordinary particles (so the product is positive) it must end with an even number of superpartners produced, so the product of the final numbers is positive. Similarly, when a superpartner decays the initial number is negative, so the final particles have to have an odd number of superpartners (usually one). This chain ends with the lightest superpartner, which has to be a stable particle because it cannot decay to all normal particles since the product of their numbers is positive.

From this analysis we arrive at one remarkable prediction of supersymmetry—there should exist a new stable particle, in addition to the electron, the up quark, and the neutrinos. Careful study implies it should be a particle that does not carry color or electric charge. Without color or electric charge it cannot interact to form hadrons or nuclei or atoms, so it cannot make dense clumps of matter that become luminous stars. It should behave as "dark matter," and indeed it has all the properties of the dark matter needed by the astrophysicists and cosmologists. The supersymmetric dark matter was not invented to provide dark matter for astrophysicists, though it indeed matches what they need. Rather, it is a natural consequence of the supersymmetric Standard theory, first noticed in the early 1980s.

In a supersymmetric theory the problem of widely separated scales described in the previous chapter is solved automatically, in a technical way. The quantum theory rules require that bosons and fermions contribute with opposite sign to any quantity that plays a role in connecting the two scales, so if every boson comes with a fermion their net contribution adds up to zero and the problem disappears, so long as the superpartner masses are not very much larger than the particle masses. The supersymmetric theory alone does not predict that two widely separated scales should exist, but it allows construction of a meaningful theory if there are two widely separated scales (as there apparently are).

The way the supersymmetric theory deals with the Higgs physics is subtle. It is a great improvement compared to the Standard Theory, but not a complete solution. The Higgs fields, and therefore the Higgs bosons, have to be included explicitly, just as all the quark and lepton fields do. But from that stage on all the needed Higgs physics can be derived, rather than requiring special assumptions as in the Standard Theory. The supersymmetric Standard Theory is presumed to have its fundamental form at the shortest distance scale. Then the extrapolation technique described in the previous chapter is used to deduce how the theory behaves at the distances probed in experiments today, and larger distances. What happens is remarkable and elegant. The conditions necessary for the Higgs mechanism to work, so that the W and Z bosons and the quarks and leptons can be given mass while the theory remains meaningful, emerge from the equations of the supersymmetric theory. As described in Chapter 8, the lowest energy state becomes the state where the Higgs field has a constant value different from zero, rather than vanishing like all the others. The Higgs mechanism is derived.

These Higgs physics results were obtained in the early 1980s, and they depended on one condition. In order for the supersymmetric theory to predict the Higgs mechanism, it was necessary for the top quark to be very heavy, heavier than the W boson if the Higgs mechanism were to work regardless of other parameters of the theory. At the time the heaviest quark known was the bottom, weighing about one twentieth of the Z, and it was hard to believe the top could be so heavy. Today we know from LEP and Fermilab data that the top is indeed heavy enough, just as required by the supersymmetric theory. Higgs physics as viewed from the Standard Theory perspective is obscure and mysterious. Much has been written about it, and many physicists are uneasy about it. If the supersymmetric version had appeared first, I think Higgs physics would have been accepted as an attractive and natural result, accompanied by little of the mystery.

## *Detecting Supersymmetry*

In the supersymmetric Standard Theory there is no alternative to the Higgs bosons existing. If they are not found, we will know that the entire theory is wrong. Today's understanding of the supersymmetric theory is not good enough to predict the Higgs boson mass, but it is good enough to prove that there is an upper limit on it, and in addition, that the numerical value of the limit can be calculated. The latter point is important—it could have happened (but it did not) that a limit could be set, but that its value depended on other quantities whose values were not known, in which case the limit would have been of no practical interest. The Higgs boson mass is expected to be above (about) the W boson mass and below (about) twice the W boson mass. This range of values could be explored at LEP or FNAL if either were appropriately upgraded, and at the planned CERN hadron collider LHC (if very good detectors are built), and at the proposed next electron linear collider, called NLC, which will be modeled on the current SLC at SLAC.

Let us turn briefly to the superpartners, the new particles associated with the ones we know. In keeping with the whimsical approach to names, the rules for naming the superpartners are simple. The partners of the bosons are suffixed—ino (photino, zino, wino, gluino, higgsino), though they are not little or light as that suffix suggests. The partners of the fermions have an s- prefix, s— (selectron, smuon, slepton, squark, stop, etc.). Supersymmetry has its own slanguage.

Should the superpartners already have been found? Is it true, as some people have suggested, that not finding the superpartners implies doubts about the validity of the theory? Or are the superpartners too heavy to be found? As was the case for the Standard Theory, we still lack the principles to predict the masses completely. But, in the supersymmetric case there are constraints that help answer these questions. The only way known to study the problem is by building models that satisfy all the theoretical and experimental constraints, and asking what superpartner masses arise in the models. The implication of building such models is that if we had been very lucky, superpartners might have already been detected, but their natural mass range is in the region that will be partly studied at LEP and Fermilab beginning in the middle 1990s, and beyond that region. Almost the entire range of expected masses is in the region that will be covered at the CERN LHC, and a substantial portion of the range could be covered at NLC if it is built. Important constraints on the masses are that if the squark and slepton and wino masses are too small or too large, the predicted amount of dark matter will not be right, the way the Higgs mechanism works will not be right, or the unification of the forces at short distances will not work out.

Perhaps the first superpartner will be found at LEP or Fermilab in the next few years. Even if that happens, only one or a few superpartners could be studied there, because LEP and Fermilab do not have sufficient energy or luminosity to produce and detect most of the superpartners. To fully sort out the structure of the theory it will be essential to measure the properties of almost all of the superpartners. Eventually it should be possible with combined information from LHC and NLC. The superpartners would leave definite, unique signatures in the detectors, so it will be possible to detect and study them if they are produced. It will not be simple to detect them—sometimes ordinary particles can mimic supersymmetric signatures—but it is always possible to learn to distinguish superpartners from ordinary particles.

If supersymmetry is discovered there may be a great bonus. If the approximate extrapolation method used to relate the distance scales probed in today's experiments to the very short distances works as well as it seems to, it essentially means that we can "do experiments" at $10^{-30}$ cm even though we cannot build a real apparatus to do that. We can formulate the theory at $10^{-30}$ cm and then calculate the predictions at $10^{-16}$ cm and test them. Or we can make measurements at $10^{-16}$ cm and then extrapolate to the form of the implied theory at $10^{-30}$ cm. In the next chapter we'll look at possible ultimate theories and see that it can be argued that a distance scale of $10^{-33}$ cm may be special for constructing a fundamental theory. That seems impossibly far from $10^{-16}$ cm, but not so far from $10^{-30}$ cm (see Fig. 9.1 on page 126).

While supersymmetric theories go a long way toward improving understanding, making the first inroads on the Input and Mechanism Understanding, they still leave much to be done. They do not help solve the "family problem"—why are there three families, not just one, not more?—at all as far as we can see. There is no understanding of how the supersymmetry is broken so the superpartners do not have the same masses as the partners. The supersymmetric extension of the Standard Theory is not the last stage in extending the Standard Theory, but perhaps it is part of it. As is usual in physics, progress will probably be made one step at a time.

# 11

# How Much Unification? Is There a Limit to Understanding?

Unification provides a valuable perspective for viewing both the past and the future of science—physics in particular. Most recently, as described in Chapter 4, we have learned that the weak and the electromagnetic interactions are unified. They look different to us, but that is because of the context in which we see them. What further unification can we hope for in the future?

## *Grand Unification*

In the previous two chapters we saw that there were good indications that indeed the electromagnetic and weak and strong forces all become the same at very short distances or (equivalently) at very high energies. That unification of the three forces has been called a "Grand Unification," to distinguish it from the "unification" of weak and electromagnetic forces. While the name is a little overstated, it is the standard terminology in particle physics, so I will use it, though in lowercase letters. We also saw in Chapters 9 and 10 that for such a grand unified theory to be consistent some new feature had to enter in order to not have the two widely separated scales collapse together, and supersymmetry does that job well. A grand unified theory and a supersymmetric one are not alternative approaches; rather, a grand unified theory that uses the Higgs mechanism needs to be a supersymmetric theory as well. Thus it is certainly fair to hope for a supersymmetric grand unified theory to emerge in the foreseeable future. As with every previous unification, our understanding of nature would be greatly improved by such a theory even though there would still be unsolved problems—such as the origin of families, the origin of the breaking of supersymmetry, the connections to gravity, and so on. If a successful grand unified theory is ever found, perhaps some unsolved problems will get solved, just as Maxwell's unification of electricity and magnetism also explained what light was.

Sometimes people argue that there are too many particles (leptons plus quarks of different colors) for them all to be fundamental, and supersymmetry even doubles the number. But if they are unified we should not think of them as different particles. We do not think of electrons whose spin points in different directions as different particles, even though they have different interactions. Similarly, we should think only of one up quark even though it can be in three different color states, and we should think of electrons and their neutrinos as different members of one doublet. If quarks and leptons are really unified then there is one basic object—let us call it a "quarton"—that looks like a quark or a lepton from different views. Then there is only one quarton for family 1—different views of it can be a red up quark or an electron or a blue down quark. There is another quarton for family 2, and a third for family 3—so far we do not understand how to unify the families. Antiparticles also should not be counted as separate, because particles and antiparticles must always exist together. Similarly, if supersymmetry is really a symmetry of nature the superpartners should not be counted as separate—they too are just a different view of the basic "superquartons."

## *Can Grand Unified Theories Be Tested?*

The grand unification of the forces takes place at a distance scale of about $10^{-30}$ cm. To clarify what that means, suppose that the colliders really were microscopes. Then when we focus to observe objects of size about $10^{-14}$ cm, we can clearly see differences between the effects of the strong, electromagnetic, and weak forces. If we focus to a size over 100 times smaller, below $10^{-16}$ cm, then we can no longer tell the weak and electromagnetic forces apart so well—they are becoming unified. When we focus to a much smaller size, all the way to $10^{-30}$ cm, we no longer can tell any of these forces from one another; the grand unification has occurred. Unfortunately, no real collider could ever probe such small distances.

That has led some people to be concerned or even to argue that grand unified theories will never be tested. While not every kind of grand unified theory may be completely testable, the particular kind described just above is definitely very testable, for two types of reasons. First, the quantum theory extrapolation method described in Chapter 9 means that the predictions can be formulated at the short distance where the full unification holds, and then calculated at longer distances. An example of this that we saw earlier is the way the electroweak unification angle $\theta_W$ is calculated—its predicted value at the short distances is 37.76°, it extrapolates to 28.8° at distances that experiments can probe, and 28.8° is what experiments measure. The masses of the superpartners (assuming they exist) also provide a large number of examples—there are about thirty-two observable masses that can be predicted in terms of as few as four parameters, so twenty-eight experimental tests can be made at colliders if all the superpartners are light enough to be detected eventually at LHC and NLC. If some are too heavy to detect at these facilities, perhaps we can still learn most of what we need to know by other methods.

Second, there are several phenomena that are predicted to occur in grand unified theories. Without the theory no one would have guessed how to interpret them had they been observed, but with the theory they can be seen as important tests. One is determining how protons decay (the experiments were described in Chapter 5), if they indeed do decay. Any theory that relates quarks and leptons will contain the possibility of transitions of quarks into leptons. Then quarks and leptons will come to be thought of as different forms of a single underlying object—they will be thought of as quartons. In some forms of the theory, this will make the proton into an unstable particle. Whether the proton decays and, if so, how rapidly, would give us significant information. In addition, what it decays into is very important. For example, in a grand unified theory based on the Standard Theory the dominant decay should be the transition of a proton into a pion plus a positron, while in one based on a supersymmetric Standard Theory the dominant decay should be to a kaon plus a neutrino (some hadrons have to occur in a proton decay because the quarks have to appear as constituents of hadrons rather than separately). So there are clear experimental ways to distinguish one grand unified theory from another.

A further important and direct way to get experimental information about the unification scale physics comes from neutrino masses. In the Standard Theory itself neutrinos are massless, but in grand unified theories they usually are not massless. Thus the pattern of neutrino masses that is finally found will help to tell us how to extend the Standard Theory into a grand unified theory. Each form of grand unified theory will predict a particular pattern of neutrino masses. In principle there are nine observables associated with the neutrino system, so a great deal of information will eventually be available, though it is not guaranteed that all nine can be measured in practice. Similar remarks hold for the masses and other properties of the quark and charged lepton systems.

Additional tests may exist, depending on the form the theory

*"But don't you see, Gershon—if the particle is too small and too short-lived to detect, we can't just take it on faith that you've discovered it."*

takes. The grand unified theory affects the baryon asymmetry of the universe (Chapter 12) and various other aspects of cosmology. And finally, as has always happened in the past, once the theory is written and taken seriously and explored by theorists, new predictions and connections to other parts of physics that provide further tests are likely to be found. One example is the prediction of Einstein, based on his general relativity theory, that light would feel the gravitational force; thus light from distant stars passing near the sun should be bent toward the sun. Observations confirmed that indeed happens. Another example is that two of the main tests of the big bang origin of the universe (the abundance of helium in the universe and the photons left over from the early expansion—see the next chapter) could not have been anticipated before the theory was worked out.

## *The Primary Theory*

In 1984 a new and ambitious approach to understanding the natural world emerged. Even a grand unification is a step in a natural progression, a continuation of the same research program that has been under way at least for three centuries. The new approach of recent years continues the traditional one in that unifying the other forces with gravity is one of its goals, but it goes further. The advocates of the new approach hope to explain *why* there are quarks and leptons and not some other particles, and to show how space and time arise and even why there are three space dimensions. They hope the new theory has no masses, force strengths, or other numbers or constants that are undetermined. In the language of Chapter 7, the goal is a Why Understanding. If such a theory is discovered, it will mean that we are at the end of the search for the underlying laws of nature.

Many of the people working in that area believe that the theory has actually been formulated, but that no one has yet figured out whether it describes our universe. The difficulty is that the theory provides equations, but our universe is supposed to be a solution to the equations. Even if the equations have been guessed, to solve them requires many stages, such as learning how space and time are defined. Even worse, basic equations normally have a high level of symmetry among the basic particles, but the solutions do not. This is the phenomenon called "spontaneous symmetry breaking," discussed briefly in Chapters 7 and 8. The reader might want to recall the example (from Chapter 7) of the equation that treated the electron, muon, and tau symmetrically, but had solutions that gave them different properties. The goal of the research of a number of theorists today is to learn whether there is a solution that describes our universe.

The ambitious theories—perhaps the plural is appropriate until one theory is selected—are often called the "Theory of Everything" by some physicists, and even more often by many journalists and writers. That name is in one sense appropriate because such a theory would explain the underlying structure that all higher levels are built on. In another sense the name is not appropriate because it does not explain things that are important to us, such as when and where earthquakes occur, or the feelings that listening to the music of Bach imparts. Here I will call it the primary theory.

Throughout this book I have been careful to say that quarks and leptons were pointlike rather than "points." These ambitious theories are based on objects called "superstrings." They represent the basic constituents (quarks and leptons) as "strings" rather than points, but that distinction would only appear at distances of about $10^{-32}$ or $10^{-33}$ cm, far smaller than any collider could probe, and 100 to 1000 times smaller than the distances where the unification of the forces seemed to occur. Again, this has led people to argue that the theory cannot be tested, that theoretical physicists are abandoning science for philosophy or mysticism.

## *Can the Primary Theory Be Tested?*

That criticism is not justified. Earlier I explained that theory had been ahead of experiment from the 1970s until now, and for the foreseeable future. But the situation is reversed in the search for the primary theory. In a sense, experiment is well ahead already. While I expect that once an explicit candidate for the primary theory is developed it will make new and unexpected predictions that provide further tests, there are already many tests it must pass. It must require quarks and leptons to exist, but not particles of similar mass that have the combined characteristics of quarks and leptons, since the latter are not observed. It must explain why we live in three space dimensions. It must explain what spin and electric, and color charge are and why some particles have them. It must predict the masses of all twelve quarks and leptons, and explain why they occur in three families. It must predict several observable properties of the universe—its total energy (consistent with zero), its total angular momentum (consistent with zero), etc. Any theory that can explain and predict those things will clearly be the one that has been searched for since the beginning, and accepting it as that will be well founded in the traditional senses of physics.

To understand that these are indeed experimental tests one must understand what "explain" and "predict" mean. How can the primary theory "predict" there are three spatial dimensions if we already know that? Or how can a grand unified supersymmetric theory predict the electroweak unification angle $\theta_W = 28.8°$ if $\theta_W$ is measured to be 28.8° before the grand unified supersymmetric theory is discovered? The reason is that once the theory is written down, the number of spatial dimensions and the values of many observables such as $\theta_W$ are uniquely determined for that theory. In general there are really three possibilities: a theory can uniquely determine a value for something that is already measured, such as $\theta_W$, and get it right, in which case it is appropriate to say the theory predicts it. Or it can determine a quantity and get it wrong, thus showing that theory is incorrect. Or it can allow a range of values for a quantity, depending on what value some other (not completely known) quantity takes, with the correct value included in the range. In the latter case we do not say the result is predicted, but that the theory is consistent with the measurement, or "could explain" the measurement. The situation is similar with "explain." For example, it

*"But we just don't have the technology to carry it out."*

is appropriate to say that Newton's laws explained Kepler's laws of planetary motion, because once Newton's laws were written it was possible to uniquely derive Kepler's laws.

A few paragraphs ago I said that the stringlike rather than pointlike nature of the particles was expected to appear only when they are probed at distances like $10^{-32}$ or $10^{-33}$ cm (if at all). There are three hints that this is the appropriate distance scale for the primary theory. First, we expect the primary theory to unify gravity with all the other forces. When the strengths of the forces are extrapolated by the method described in Chapter 9, they do indeed all (including gravity) become similar, at a distance scale of about $10^{-33}$ cm. That is, if we could see down to interactions occurring at that size, the effects of gravity would be as large as the effects of the other forces (which already had become similar to each other by about $10^{-30}$ cm). Second, a quantum theory of gravity has not yet been formulated. It is possible to ask if it matters whether there is a quantum theory of gravity. The answer is that the results of Einsteinian gravity should not need modification as systems get smaller, until we consider sizes of about $10^{-33}$ or smaller. Third, if one asks what is a natural distance for formulating the basic laws of nature, not paying any attention to the presence of people or intermediate structures such as atoms, then one has only the fundamental constants such as Planck's constant **h**, Newton's constant **G**, and the speed of light **c** to work with. From them only one quantity can be formed with the units of length, the square root of $Gh/c^3$, with a numerical value of about $10^{-35}$ cm. This way of thinking is not one we are used to in everyday life, but some reflection on the meaning of having a natural size occur suggests that the basic laws of nature should be formulated at about that distance, so that the natural size enters in a simple way. These numbers that come from different approaches are all about the same. All of these arguments suggest that the primary theory should take its simplest form at a distance scale of about $10^{-33}$ cm. This scale is called the "Planck scale," after Max Planck, who was the first to make such arguments, early in the twentieth century. In a sense the Planck distance is the natural size for the universe. What needs explaining is why things our size, thirty-five powers of 10 larger, should exist.

That distance is only about 100 times smaller than the scale where the weak, electromagnetic, and strong forces unify, which leads theorists to expect an additional set of experimentally testable predictions will be found. Though at present the theory is too poorly understood to give a compelling example, it is possible to expect that because of the nearness of the Planck scale to the unification scale, there will be several modifications to predictions for proton decay properties, neutrino masses and even $\theta_W$; these are presently being studied.

## *Are There Limits to Understanding?*

I hope the discussion of the last few pages has established that it will indeed be possible to test candidates for the primary theory by the traditional standards of physics even though the theory is expected to take its simplest form at distances that are only probed indirectly with experimental data. There is another question—perhaps for some reason we will not be able to formulate such a theory. Perhaps science has limits and we can only describe how nature works but not figure out why it is the way it is. Several different reasons have been given to support the claim that we will not be able to find a primary theory—to reach a Why Understanding, in the terminology of Chapter 7. Similar reasons are sometimes given to claim that humans will not be able to understand human consciousness. Earlier I commented briefly (Chapters 7 and 9) that I am so far unconvinced by any argument I have seen to the effect that science has limits, or that we cannot construct a primary theory, or achieve a Why Understanding. Of course the absence of an argument against such success does not imply that we will succeed.

We won't know whether humans can find a Why Understanding until a primary theory is worked out, which would settle the question one way, or until everyone gives up trying, which would settle it the other way. So far in science it has worked to keep trying, and that will go on for a long time regardless of the arguments on either side.

# 12

# Particle Physics, Astrophysics, and Cosmology

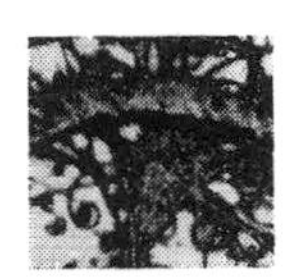

Traditionally, particle physicists have studied the smallest things in the universe, while cosmologists have studied the largest—including the universe itself. Astrophysicists study the physics of astronomical objects, stars, galaxies, etc. Astronomers observe and analyze the astronomical phenomena. Anyone can wear several of these hats. The phrase "astroparticle physics" is already in use, and probably we should expect "cosmophysics" soon. This chapter is a glimpse at how developments in particle physics, astronomy, astrophysics, and cosmology are affecting one another, particularly in trying to understand the big bang, dark matter, and why the universe has more matter than antimatter.

## *The Big Bang*

It is now widely accepted that our universe began with a "big bang" about 15 billion years ago. According to this theory, all matter, and space itself, initially occupied a tiny region. Since the big bang the universe has been expanding. There are several strong pieces of evidence for this theory, which has become increasingly well established as more observations are made.

First, that the universe is expanding is directly observed—in whatever direction an astronomer looks, galaxies are moving apart. Since light travels at a fixed speed, the light we see from distant stars began its journey long ago, and information from that light tells us that the universe was expanding then too.

Just after the big bang, the universe was a dense gas of particles of all kinds (all the quarks and leptons and bosons and any other particles provided by extending the Standard Theory) moving around randomly, colliding, undergoing all possible reactions. As expansion continued, the universe cooled. That is a familiar thing—if you use a paint spray can, the compressed gas inside expands and the can becomes colder. Conversely, if you pump air into a tire, the air is compressed and the tire becomes warmer. After a long time most of the original particles (W's, top quarks, etc.), being unstable, had decayed. At first there was enough energy when collisions occurred to make more W's, top quarks, etc., but as cooling continued, the particles had less energy, and no more heavy unstable particles were produced. Eventually none remained.

A little later it was cool enough so that quarks could combine to make protons and neutrons. Some protons and neutrons combined to make deuterium, and some deuterium nuclei combined to make helium nuclei (each with two protons and two neutrons). Calculations show that after a while the nuclei of the universe were 93 percent hydrogen and 6 percent helium. In addition there were lots of photons and neutrinos resulting from the annihilations and decays of other particles.

This history of the universe leads to two quantitative predictions that are accurately confirmed by observation. First, the observed amount of helium, called the "helium abundance," is, to good accuracy, equal to the predicted amount. Second, the expected energies of the remaining photons can be precisely calculated from this history, and the observed energies agree with the calculated ones to a fraction of a percent. The jargon for this is that the remaining photons are the "microwave background radiation," and the observed energies of the photons are their "temperature." The big bang history and its associated observations, are very clearly described by Steven Weinberg in his book *The First Three Minutes*.

According to Einstein's theory of gravitation, energy and mass are related, so anything with energy will attract other things gravitationally. Neutrinos left over from the decaying particles provide a gravitational attraction for all particles. More gravitational attraction slows down the expansion, so the number of different kinds of neutrinos affects how rapidly the universe is predicted to expand. Changing the expected expansion rate would have changed the prediction for the percent of nuclei that were helium. The calculation can be done as if there were two, or three, or four, or more kinds of neutrinos, and the prediction then compared with the observed fraction of helium. The three known neutrinos gave the right answer—two or four neutrinos would have given a prediction that disagreed with observation. When Z decays were studied at LEP (starting in 1989) it was possible to deduce how often a Z was created and then decayed into neutrinos, even though the neutrinos escape the detector. The results show that the three known neutrinos, $\nu_e$, $\nu_\mu$, $\nu_\tau$, precisely account for the LEP observations. That confirms the validity of the cosmology arguments and reinforces our confidence in the description of the big bang view of the universe.

The earliest moments of the big bang history may offer scientists an opportunity to learn more about particle physics regimes that are not directly accessible by experiment. Immediately after the big bang, at $10^{-43}$ seconds, the particle energies were so large that, effectively, the big bang itself provided a microscope down to distances of the sizes of the hypothetical superstrings mentioned in Chapter 11; perhaps some consequences of the superstring interactions can be deduced that could be tested by observations. A little later ($10^{-40}$ seconds) the scale of the grand unified theories was reached. Until that time a number of heavy bosons associated with the unified force would interact with the Standard Theory particles. As the universe cooled below that stage, those bosons decayed away and no more were produced, but possibly their brief presence would affect something we can observe today.

## *Dark Matter*

Another major intersection of particle physics with cosmology and astrophysics is the topic of dark matter (Chapters 9 and 10). This is a fascinating area of current research. If the universe contains a lot of matter, the mutual gravitational attraction of all the matter will eventually slow down the expansion and the universe will collapse back on itself. If the universe contains very little matter, it will keep on expanding. There is a precise amount between these two alternatives for which the expansion will slow down but never finally turn around into a collapse. This is sometimes called "just closing the universe," or a "flat universe."

Some part of the total matter in the universe is visible, of course, in stars, and more occurs as "dust" clouds that are illuminated by starlight. We used to assume that all the matter would be in these forms, but now we think some is dark, invisible to traditional observations except for its gravitational effects. To talk about the amounts of different kinds of matter it is usual to define the ratio of a given kind of matter to the amount needed to just close the universe. The symbol usually used to denote this ratio is $\Omega$ (capital Greek omega). If the universe is just closed, then $\Omega_{TOTAL} = 1$. (The symbol $\Omega_X$ could stand for any kind of matter in the universe today; for example, X = VIS for visible matter, or X = SUSY for supersymmetric partners, or X = TOTAL for all forms of matter added up. The value of $\Omega_X$ is calculated by taking the amount of matter of form X and dividing it by the amount of matter just needed to close the universe.)

Many cosmologists and particle physicists expect, because of theoretical arguments, that $\Omega_{TOTAL} = 1$. Astronomers measure the amounts of matter various ways. For example, watching how a galaxy moves allows astronomers to deduce how much gravitational attraction the galaxy feels from other matter. They find results typically in the range $\Omega = 0.2$ to $0.6$, with big uncertainties. Observations in progress, particularly with the Hubble space telescope, will greatly improve our knowledge of what $\Omega$ is, and whether $\Omega = 1$.

The subject gets exciting when the amount of matter made of baryons is worked out. Including all relevant observations and indirect constraints, $\Omega_{\text{BARYON}} = 0.05 \pm 0.02$. This is much too small to account for what is observed other ways, or what most cosmologists expect. It includes baryonic matter of all forms, based on direct and indirect evidence. The visible part of the matter is in the form of stars, like our sun, made of the same protons and neutrons and electrons we are made of, and interstellar "dust" that emits x-rays when its atoms are excited. If $\Omega_{\text{TOTAL}}$ really is 1 then most of the matter of the universe is not made of quarks and leptons. It does not make stars. It is "dark matter." If $\Omega_{\text{TOTAL}} = 1$, most of the universe is made of kinds of particles that we are not made of.

Particle physics is relevant to the study of dark matter because extensions of the Standard Theory predict dark matter. We have encountered two possibilities already, massive neutrinos and supersymmetric partners. If neutrinos have mass they will contribute to the dark matter. There is not yet any compelling argument about how much mass neutrinos have, but crude particle physics models can give the tau neutrino a mass that puts $\Omega_\nu$ in the range of 0.1–0.5. If the tau neutrino has such a mass, experiments underway at CERN and Fermilab are likely to measure the mass of the tau neutrino. Even more remarkable in my view is that the lightest supersymmetric partner (see Chapter 10) is a good candidate for dark matter. If supersymmetric partners are included in the dense gas of particles generated by the big bang, the number of them left as the universe cooled can be calculated, and they naturally give $\Omega_{\text{SUSY}} = 0.5–1$. Dark matter emerged as a major prediction from supersymmetry studies in the early 1980s, well before nonbaryonic dark matter was established as a clear observational result from astronomy. The particle physics expectation that dark matter should exist would have been made with or without the astrophysical observation. If superpartners are detected at colliders, it will be possible to measure the mass of the lightest superpartner, the dark matter particle, and then to calculate $\Omega_{SUSY}$.

Even better, unified particle theories typically automatically have both massive neutrinos and superpartners, so they predict a mixture of the two forms of dark matter. Since the neutrinos are rather light they move rapidly and are called "hot dark matter." The lightest superpartners are more massive and move slowly—they are "cold dark matter." So far no theory is sufficiently well understood to predict the relative amounts of hot and cold dark matter, but a number of physicists are trying to improve our understanding and make such a prediction. In recent years two different results from astrophysics indicate that a mixture of about two-thirds cold dark matter and one-third hot dark matter may do a good job of describing what is observed. One set of results comes from observations of microwave background radiation recorded by the COBE satellite, and the other comes from studies of how the formation of structures such as galaxies and clusters of galaxies can best be described. Understanding dark matter is one of the central scientific problems of our time, and it necessarily requires a combination of particle physics, astrophysics, and cosmology.

## *The "Baryon Asymmetry"*

Recall from Chapter 4 that quarks and gluons combine into hadrons of two kinds, mesons made of quark and antiquark, and baryons (including the proton and neutron) made of three quarks. All mesons, and all baryons except the proton and neutron are unstable and decay quickly. (Free neutrons also decay, very slowly, but for subtle reasons, neutrons in nuclei do not decay.) So our world is made of baryons and electrons.

Just as three quarks make a baryon, three antiquarks make an antibaryon. Since the antiparticles are just particles too, we have no known reason to expect more baryons than antibaryons in the universe. If the universe actually had an equal number of baryons and antibaryons we could imagine that they were created from concentrations of energy (as is done at colliders), and then somehow they separated. On the other hand, if the universe actually were made only of baryons, as it seems to be, we would be left wondering how they got there—there seems to be no mechanism to give only baryons.

Could there be equal numbers of baryons and antibaryons even if there are no antibaryons in our part of the universe? The evidence seems to indicate otherwise. Since matter in most of the universe is in the form of electrically neutral atoms, if there were equal numbers of protons and antiprotons, there would also be equal numbers of electrons and positrons (antielectrons). Galaxies collide with other galaxies often, and whenever a matter galaxy collided with an antimatter galaxy, lots of electrons and positrons would annihilate. Under those conditions, they would produce lots of photons of a definite known energy, so experiments continue to look for those photons. They have not been found. Careful analysis of such arguments suggests fairly strongly that our universe is all or nearly all matter rather than an equal mixture of matter and antimatter.

Initially that result seemed to imply that we could not hope to understand the origin of the matter in our universe. However, in 1967 the Russian theoretical physicist (and dissident and human rights advocate) Andrei Sakharov pointed out that if several conditions were satisfied by the laws that described nature, then it was possible to start out at the big bang with equal numbers of baryons and antibaryons, and end up with a universe made mostly of baryons. Most of Sakharov's conditions are requirements that the particle physics must satisfy. Although he used different language to describe it because his work came before the Standard Theory, his most speculative condition required having an interaction that could turn quarks into leptons. In fact, the grand unified theories described in the previous chapter do just that. Even after the grand unified theories were written, it still took several years before people realized that they had the potential to explain the observed baryon asymmetry of the universe. Then a number of theorists constructed models to see if the right answer would emerge. We still do not know the outcome—this is an active research area. It is possible to make models where the right answer does emerge, but so far only by making some assumptions that are hard to check.

Most of the work on the baryon asymmetry problem falls in the area of particle physics, but one big uncertainty is a thorough mixture of particle physics and cosmology. When water cools, at some stage it makes what is called a "phase transition"; it turns into ice. The universe also undergoes phase transitions as it cools. Understanding the details of those phase transitions may be required to derive the baryon asymmetry.

Experiments searching for proton decay also affect the baryon asymmetry question, since a proton can only decay by changing some of its quarks into leptons. If protons are observed to decay (some experiments were described in Chapter 5) we will have direct experimental proof that such an interaction exists and will be able to deduce whether it has the right properties to produce the baryon asymmetry. Any explanation of the baryon asymmetry must include a proof that it generates a large enough baryon asymmetry without contradicting proton decay data.

## *Inflation and the Origins of Large-Scale Structure*

Two main topics of current research in cosmology may depend on the underlying particle physics. For several reasons, the expansion rate of the very early universe is thought to be highly non-uniform, with a short period of very rapid expansion followed by an abrupt slowing to the present rate. The rapid expansion is called the "inflationary universe." The equations that describe it depend on a knowledge of all of the Higgs boson and related fields that were present as the inflation began, so the theory of the inflationary universe cannot be settled until the relevant particle physics is known.

The structures we see in the universe today are thought to have formed simply by gravitational attraction. If small concentrations of energy developed as the hot, dense gas of particles expanded, other particles were attracted to the more massive concentrations. Denser regions eventually turned into clusters of galaxies, less dense regions into voids. The crucial question is what determines the initial density fluctuations and concentrations; that is where the particle physics enters. In order to produce concentrations that agree with cosmological conditions, the particle physics theory must satisfy several constraints, which provide important tests.

## *Cosmoastroparticle Physics*

The unification of the smallest and the largest things in the universe, perhaps including the largest thing—the universe itself—is one of the telling trends of recent years in physics. Each field found it needed the others. Perhaps collaborations among cosmologists and astrophysicists and particle physicists will help us find the answers. Perhaps young cosmoastroparticle physicists, trained in all three areas, will be needed to finish the integration.

# 13
# Understanding a Flower

Imagine particle physics has been understood as well as possible, even to the level of a Why Understanding. What does that tell us about a flower—its colors, its intricacy, how it unfolds and grows, or dies? Almost nothing. Knowing how atoms work explains why molecules form, and lets us understand which molecules are more stable than others. But because of the incredible complexity of the behavior of the huge numbers of molecules needed for anything such as life to happen, we can never predict that a self-replicating molecule will exist—or not exist. We can never predict from knowing only about quarks and leptons and their interactions that flowers will exist. It is not that if we were a little smarter then we could make such predictions; the complexity and the number of alternatives are so overwhelmingly large that there would not be time in the lifetime of the universe to deduce that red roses would exist. Some of the huge number of possibilities will happen; most will not.

The opposite is not true. Once we know that they exist, we can attain a Descriptive Understanding of flowers. We can understand their colors in terms of the molecules that make them up and the properties of sunlight, we can understand their genetic material, and each stage down to their constituent atoms. Once we analyze a flower below the level of its genetic material, we have left the domain of flowers and entered the domain of molecular biology. We can marvel at a flower for its beauty, and its scent, and also because we can understand it scientifically as well. As Richard Feynman wrote in the *Feynman Lectures on Physics* in 1963, "What men are poets who can speak of Jupiter if he were like a man, but if he is an immense spinning sphere of methane and ammonia must be silent?"

In the terminology of Chapter 7, we can have a Descriptive Understanding of a flower, but not an Input and Mechanism or a Why Understanding. To understand a flower we must take the molecules that form its genetic material as given—once we take them apart we get a small set of molecules that make up all living matter, and no trace of a flower is left. The process continues—if we take the molecules apart we get to the ninety-two chemical elements, and no trace of the huge diversity of molecules remains. We can have a Descriptive Understanding of molecules, but the existence of the atoms is just something that is there for the molecules; if you work at the molecule level, the atoms are irreducible, basic elements. If you stay at the level of the flower, whether molecules have structure (combinations of atoms) is not only irrelevant, it is a meaningless question.

## *All Scientific Understanding So Far Is Descriptive Understanding*

Scientifically we can view nature as a hierarchy. Quarks and gluons form protons and neutrons, protons and neutrons form nuclei, nuclei and electrons form atoms, atoms form molecules, molecules form biological molecules, biological molecules form cells, cells form living organisms, some living organisms become conscious. When we turn it around and go downward we have the reductionist approach that has, over the past four centuries, been immensely productive. Science would have been impossible without it. At each level of the hierarchy we can hope to attain a Descriptive Understanding, but not more. At each level some

*"I'm on the verge of a major breakthrough, but I'm also at that point where chemistry leaves off and physics begins, so I'll have to drop the whole thing."*

building blocks and/or principles must just be accepted—the inputs and mechanisms remain unexplained in terms of that level of the hierarchy.

If all fields and subfields of science are thus aiming for a Descriptive Understanding, no field or subfield can be said to be more fundamental than any other. The understanding of particle physics achieved by the Standard Theory, the understanding of atomic physics achieved by knowing the rules of quantum theory and the theory of the electromagnetic force, the growing understanding of condensed matter physics, the understanding of the genetic code, are all Descriptive Understandings. All have essentially the same logical status.

But there is one sense in which particle physics (or, better, cosmoparticle physics) is different from all other fields. Think from the top down—biological molecules, atoms, nuclei, protons, quarks. At each stage the understanding relies on just accepting the next deeper stage. For example, in atomic physics the nuclei are just accepted as elementary objects with certain measured properties. That does not mean the next stage (or any stage) is more fundamental, since it too has its measured inputs and mechanisms. However, today particle physics aspires to attaining Input and Mechanism Understanding, and Why Understanding. That is, today some particle physics is directed toward having no inputs from outside of particle physics, and even toward understanding why each aspect of the Standard Theory is the way it is (Chapter 11). Particle physics is a little different from other fields, because particle physicists hope to achieve kinds of understanding not available to the others. It is premature to say whether such kinds of understanding will actually be achieved for particle physics—if they are, then particle physics will be more fundamental than other fields of science. Because particle physics can at least aim for a deeper understanding than other fields, there is a weak sense in which it could be said to be more basic than others; until we know that such deeper understanding can be attained, this is not a profound distinction.

## *Complexity and Chaos*

Throughout the development of physics, efforts have, appropriately, focused on the simplest systems. The equations of quantum theory can be used, along with a knowledge of the forces, to calculate how simple systems will behave. It can happen that when equations are written for a complicated system of particles, the resulting behavior of the solutions to the equations can be very different from the behavior for one or two particles. Having a Descriptive Understanding of particle physics does not imply at all having an understanding of superconductivity or of a flower, because many particles get involved and entirely different results can emerge. Because of that, many systems we meet in the everyday world have been inaccessible to a Descriptive Understanding, systems such as the weather, the turbulent flow of water, and many more.

Largely since the early 1970s, an exciting new subfield of physics has developed to bring complicated systems under the purview of physics by studying the solutions of the equations that describe these complicated systems—it is called complexity theory or chaos theory. This subfield involves new techniques that physicists hope can provide understanding of many previously inaccessible phenomena of the everyday physical world. (The same techniques are, with great optimism, being applied to many biological and social and economic phenomena; I am only commenting on their application in physics.) These new approaches are not inconsistent with the Standard Theory nor do they add to its equations—they assume the rules of quantum theory and the forces of the Standard Theory. They address the solutions of the equations of the theory, not the equations themselves. Indeed, no one has yet imagined any application of complexity theory for quarks and leptons. But in atomic physics and condensed matter physics the new techniques may have many applications. It is hoped that complexity theory and chaos theory will add significantly to the phenomena for which science has a Descriptive Understanding.

## *Particle Physics and Flowers*

Understanding quarks and leptons does not enable us to deduce that a flower garden will exist, or that we will exist. Botanists understand flowers in terms of the photosynthesis process that provides energy for growth in terms of chemicals that absorb some colors of light so others are reflected to be appreciated, and so on. But why does photosynthesis occur, and why are those colors of light absorbed and not others? Chemists and atomic physicists answer that—it is because of certain properties of atoms, which they understand. But why do atoms have those properties? Because they are systems of electrons bound to nuclei by the electromagnetic force, obeying the rules of quantum theory. What are nuclei? What are electrons? Why does the electromagnetic force have the form it does? Why are the rules of quantum theory what they are? Some of these questions are answered by the Standard Theory, some may be answered by the extensions of the Standard Theory that are the subject of research today. Whatever the eventual outcome of the quest to understand flowers this way, in addition to the poets' way, the particles and their theory are near the end of the tale. The particles are the seeds for the garden that is our world.

# APPENDIX A

# What Can Happen in Nature?—Feynman Diagrams

Richard Feynman invented a way to embody the rules of quantum theory in the simple, elegant diagrams known as "Feynman diagrams." Using them, it is easy to summarize the Standard Theory pictorially. Each particle is represented by a line. For example,

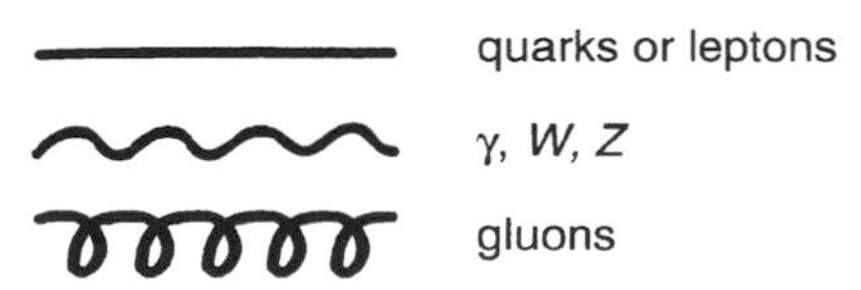


A diagram with some number of lines meeting at a point is called a vertex. There are only a few different vertices (shown in the next section). The rules are:

- Combine vertices into bigger diagrams in all possible ways.
- Every resulting diagram represents a possible process that can occur in nature.
- There is a set of mathematical rules to associate a probability of occurrence with the process represented by every diagram. In particular, each vertex has a numerical factor representing the strength of the associated force; the numerical factor measures how important that vertex is for the process considered.

- Everything that can occur in nature is represented by the diagrams you get by combining vertices.

## *The Vertices*

Here are the vertices for the weak, electromagnetic, and strong interactions of the Standard Theory.

The Electromagnetic Interaction

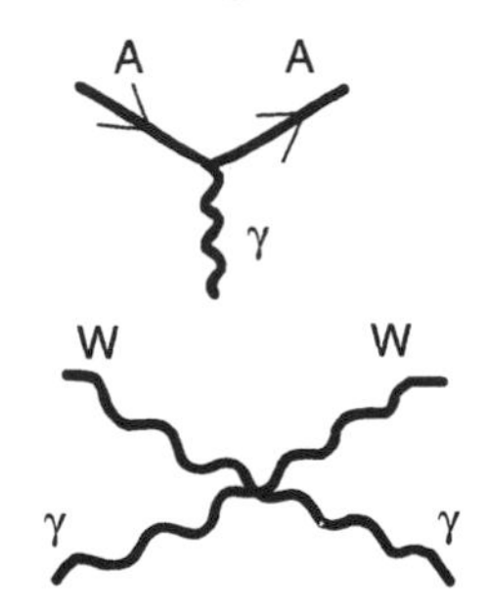


A = e, $\mu$, $\tau$, u, d, s, c, b, t, W (but not $\nu$, $\gamma$, g, Z)

The Weak Interaction

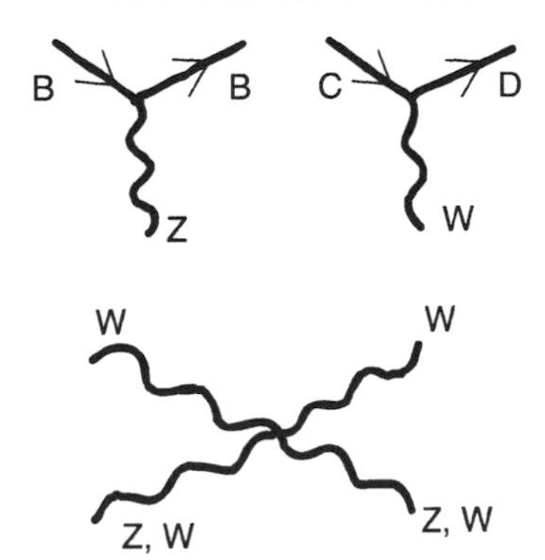


B = e, $\mu$, $\tau$, $\nu_e$, $\nu_\mu$, $\nu_\tau$, u, d, s, c, b, t, W (but not $\gamma$, g, Z)
C $\rightarrow$ D = d $\rightarrow$ u, s $\rightarrow$ c, b $\rightarrow$ t, e $\rightarrow$ $\nu_e$, $\mu \rightarrow \nu_\mu$, $\tau \rightarrow \nu_\tau$;
s $\rightarrow$ u*, s $\rightarrow$ t*, d $\rightarrow$ c*, b $\rightarrow$ c*; b $\rightarrow$ u**, d $\rightarrow$ t**

The Strong Interaction

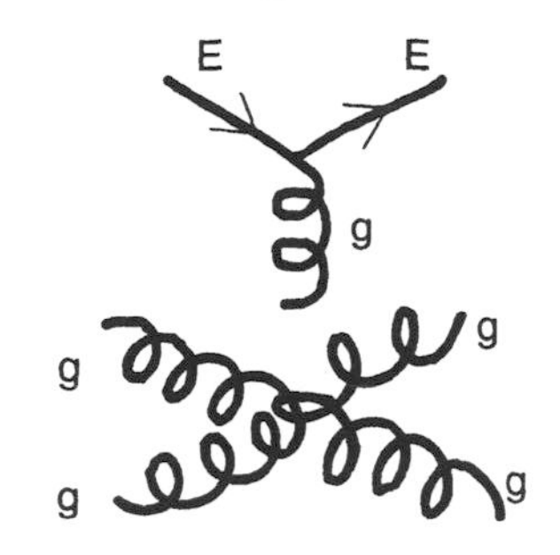


E = u, d, c, s, b, t, g
(but not $\gamma$, e, $\mu$, $\tau$, $\nu$, Z, W)

The Higgs Interation

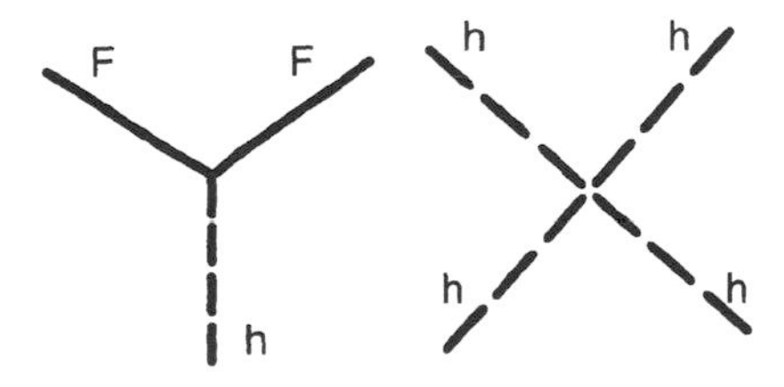


F = e, $\mu$, $\tau$, u, d, c, s, t, b, W, Z, h
(but not $\nu$, $\gamma$, g)

For all of these vertices the factor representing the force strength is a number ranging from about 0.25 to about 1.5, except for a few shown under the Weak Interaction, C → D involving W's; for those, the vertices with one * are about 20 times smaller, and the ones with ** are about 400 times smaller. The strength of the Higgs boson interaction is proportional to the mass of the interacting particle. Every vertex is accompanied by a second one for which each particle is turned into its antiparticle and all arrows are reversed. If no arrows are shown the line can go either way. When $\nu$ does not have a subscript it can be any of the neutrinos.

## Examples of Processes

For essentially all processes the probability of the process occurring is largely determined by the simplest diagrams that can be constructed. More complicated diagrams going from a given initial to a given final state can always be added, but they only modify the probability by a small amount. Here are several examples of how to make more complicated diagrams from the vertexes. Antiparticles are represented by having a bar over the particle. If no diagrams can be drawn for a process, the Standard Theory predicts that it will not occur.

THE PROCESS $\overline{\mathbf{e}} + \mathbf{e} \rightarrow + \overline{\mathbf{b}} + \mathbf{b}$

The simplest diagrams are

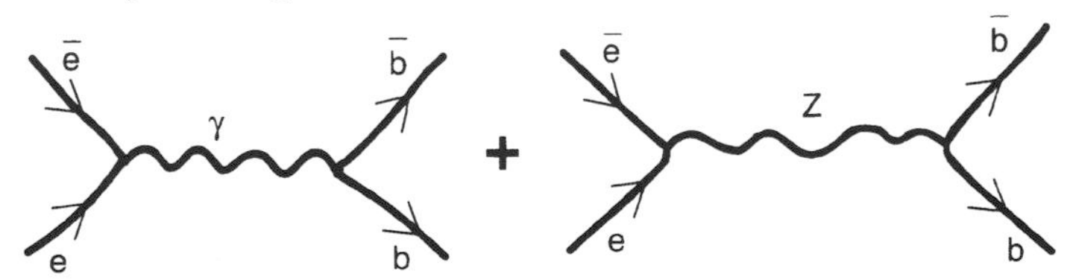


A more complicated diagram that changes the probability a few percent is

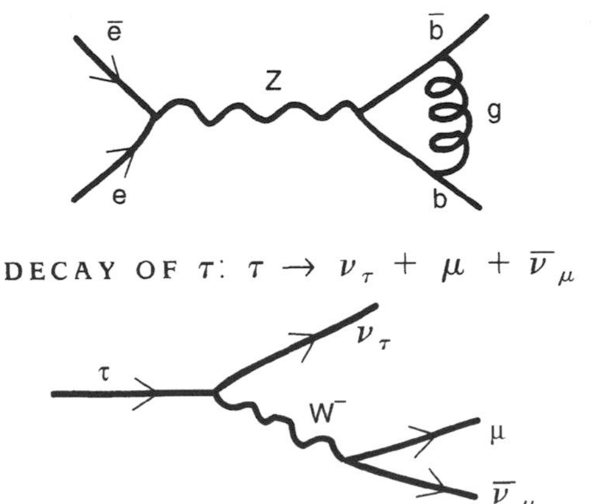


DECAY OF $\tau$: $\tau \rightarrow \nu_\tau + \mu + \overline{\nu}_\mu$

## PRODUCTION OF TOP QUARK

At the Fermilab collider, protons and antiprotons are made to collide. Sometimes a quark in the proton will collide with an antiquark in the antiproton. Then many different things can happen, corresponding to all possible diagrams that can be drawn, starting with $q + \overline{q}$; for example, $q + \overline{q} \rightarrow g + g$, $q + \overline{q} \rightarrow e + \overline{e}$, and many more. Each process has a definite probability. One process is the production of a top quark t plus an anti-top quark $\overline{t}$. The t and $\overline{t}$ are unstable so they will decay into other quarks and leptons. Here is one diagram that determines what can happen. Particles not shown decaying will emerge into the detector. This is one kind of event used to detect the top quark. This entire process takes place in about $10^{-20}$ seconds.

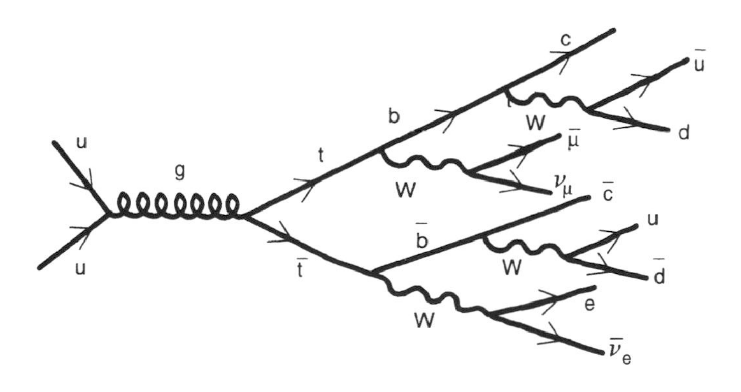


## THE DECAY $\tau \rightarrow \mu + \mu + \overline{\mu}$

No diagram, no matter how complicated, can be drawn with an initial $\tau$ and final state of two muons and an antimuon. This is one process whose occurrence would falsify the Standard Theory if it occurred with about the same probability as $\tau \rightarrow \nu_\tau + \mu + \overline{\nu}_\mu$ shown above, or help us learn how to extend the Standard Theory if it occurred with a tiny probability.

### THE DECAY b → s + γ

Here is a process that cannot occur if only "tree" diagrams are considered, where lines keep branching, because no vertex with only b and s and $\gamma$ occurs in the set at the beginning of this appendix. But it can occur when loops are included; the diagram shown is built of allowed vertexes. Diagrams with loops always have a reduced probability of occurring.

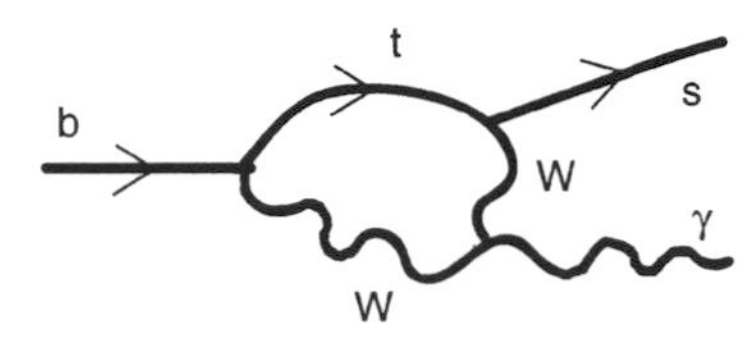


## *When the Standard Theory Is Extended*

The vertices shown are all that exist if the Standard Theory describes all of nature. Everything we see is due to these vertices, combined in many ways. Any process in biology or chemistry, if decomposed step by step into its component processes, will end up at these vertices.

If the Standard Theory is extended, more vertices will be added. If the world is supersymmetric, every vertex of the Standard Theory has added to it all possible ones with superpartners replacing particles (in pairs). If further unification of the forces occurs, some additional bosons may enter too. The additional vertices will imply some new processes should occur, or allow additional diagrams that may modify the probability of some existing processes.

# APPENDIX B
# Internal Symmetries and the Standard Theory

Since the development of quantum theory, considerations about the symmetry properties of physical systems have helped simplify and improve our understanding of those systems. Whenever a physical system is invariant under some change, or even approximately invariant, the theory that describes that system will reflect that invariance. For the Standard Theory, symmetries are even more important—they partly determine the form of the theory, as well as helping greatly to carry out and understand calculations.

Consider the hydrogen atom. In quantum theory, atoms exist in discrete energy levels. The Schrödinger equation that describes the hydrogen atom tells us that some of these energy levels occur more than once. That is a consequence of an invariance of the interaction that binds the atom under rotations in the laboratory—the energy levels of the atom do not change if you simply rotate the atom. The observed results for sets of energy levels can be derived by taking account of the invariance of the atom under spatial rotations. Masses of particles are like energy levels of atoms, so wherever sets of particles have similar masses it may be fruitful to look for invariances.

The new symmetries that underlie the relations among Standard Theory particles are not connected with transformations in space and time, but instead with changes in the identities of particles. Sometimes it is useful to think about the up and down quarks not as separate, unrelated objects, but as two different states of a single "particle," a "family-1 quark." More important, when one adds the idea that the form the theory takes cannot depend on whether the family-1 quark is up or down or in between, then certain predictions can be made, and they are confirmed by experiment. When the theory is invariant (in this example that means allowing the family-1 quark to be any combination of up and down quarks), then there is a symmetry, which often has an elegant mathematical formulation.

To understand this better, consider neutrons and protons. They have a symmetry similar to that of quarks and leptons, but one that is somewhat easier to visualize. The neutron and proton have almost the same masses (the neutron is 1.0014 times heavier than the proton); no other particles have a similar mass. Both neutrons and protons form nuclei. They interact with each other and with nuclei similarly. Why do we think of protons and neutrons as different objects? Well, obviously, because the proton has electric charge and the neutron does not.

However, nuclear interactions don't depend in any way on electric charge, and nuclear interactions are much stronger than electromagnetic interactions, so the electric charge is in some ways just a minor perturbation. Such reasoning led to the idea that neutrons and protons should really be thought of as two different states of a single "particle," a nucleon. When such ideas are formulated mathematically, if is natural to generalize them and imagine a space, "nucleon space." This space is an imaginary space, which we call "internal," in which a nucleon can point in any direction. If it points (say) up it is a proton, if it points down it is a neutron, and in some other direction it is a "mixed state." Then the electric charge is a kind of label in this space.

Generalizations such as this are often very productive in physics, but for this approach to be useful several conditions must hold. First, the form the theory takes must imply that experiments will see only the neutron and proton, but not the mixed states pointing in other directions. Second, it must make some predictions about neutron and proton interactions that are not valid when the neutrons and protons are thought of as simply separate. Third, other particles should also "fit" somehow into nucleon space.

All of these conditions hold, and nucleon space is indeed a useful way to understand the interactions of protons, neutrons, and other hadrons (for historical reasons we actually call this "strong isospin" space rather than nucleon space).

It's harder to imagine applying the internal space kind of reasoning to quarks and leptons because they seem so dissimilar, but in fact this approach is extremely successful. The mathematical formulation of the Standard Theory uses three different internal spaces. One is an imaginary circle. If a particle is moved around that circle it retains its identity, but its mathematical representation is multiplied by a factor, depending on how far around the circle the particle has gone. The requirement of invariance implies that the equations of the theory can only contain terms that are unchanged by that multiplicative factor; the form the theory can take is significantly restricted by that requirement. The mathematical language used for describing these symmetries is called group theory; the particular symmetry corresponding to invariance under these multiplicative factors is called a U(1) symmetry (the jargon is it's a "unitary group with one parameter").

The second of the three internal spaces is called a "weak isospin" space, by analogy with the "strong isospin" space of the proton and neutron. In the weak isospin space each of the pairs in vertical columns of boxes in Fig. 4.1 behave like the proton and neutron. There is a general family-1 quark that is an up quark when it points in one direction, a down quark in the opposite direction, and a mixture in other directions. There is a general family-2 lepton that is a muon neutrino when it points up, and a muon when it points down. An important requirement for this to work mathematically is that the difference of the electric charge of the "up" state minus that of the "down" state is always the same, because the electric charge is just a label in this space, and that is satisfied, the difference being 1 (in units of the proton's charge) in all cases. In the language of group theory this is an "SU(2) symmetry."

The unification of the electromagnetic and weak interactions into the electroweak theory described in Chapter 4, from this point of view, says that the "circle" internal space of the U(1) symmetry, and the "weak isospin" SU(2) space, are actually somewhat related spaces.

The third internal space is for the color labels of the particles, a three-dimensional space for each quark. If the quark points one way it is a red quark, perpendicular to that a blue quark, etc. That is an SU(3) space. The horizontal rows of boxes for the quarks in Fig. 4.1 contain the states that mix under this symmetry; the leptons have only one state and no interesting properties under this symmetry.

The requirements imposed on the mathematical form the theory is allowed to have, once one knows these symmetries should be satisfied, are very strong, nearly determining the theory. For historical reasons, the invariances of the kinds I am describing here are called "gauge invariances," and the theories determined by them are called "gauge theories." The three symmetries we know of determine the form of the Standard Theory equations. If there were further unification, in this language, there would be a larger internal space and each of the three internal spaces we have found so far would be a projection in a reduced number of "dimensions." For example, if the larger space were a sphere and you only looked at a slice through it, the slice would be a circle.

A good deal of the activity of particle theorists today in the direction of extending the Standard Theory takes the form of educated guessing about a larger internal space that might include the U(1), SU(2), and SU(3) symmetries, and then finding predictions to test the hypothesis. The mathematics of group theory sharply limits the possible spaces, so following a mixture of mathematical and experimental clues may help to produce a better description.

# APPENDIX C
# CP Violation

In the middle 1960s a small but extremely interesting effect was observed, which is called "CP violation." It may be one of the few clues we have to help solve the mystery of how to extend the Standard Theory.

"CP" is an operation that can be performed on the equations that describe particle interactions. Suppose an equation is believed to describe the behavior of electrons, including how they interact. Then an operation, called "charge conjugation" and symbolized by "C" can be applied to that equation. The operation is defined so that it converts the electron into an anti-electron (positron), and makes certain modifications to the terms in the equation (mainly it changes the sign of all charges in the equation). Experiments can be done to find out if the new equation describes the behavior of positrons. If it does, then the operation C is a symmetry of nature (and conversely). It turns out that C is observed to be a symmetry of strong and electromagnetic interactions, but not of weak interactions. In fact, for weak interactions it fails to be a symmetry not by a little (as it would have if the positrons had behaved only a little differently from electrons), but "maximally," in that some weak processes occur normally for electrons but do not occur at all for positrons. The Standard Theory incorporates all this information in a very natural way, so we can say we have a Descriptive Understanding of it.

Another operation, called "parity" and symbolized by "P"

can also be applied. P reverses the signs of all coordinates, and involves additional sign changes for some particles. Again, one finds a new equation and again experiments show that P is a symmetry of strong and electromagnetic interactions, but not weak interactions. Again, weak interactions violate parity "maximally"; some processes that can occur for a system do not occur at all for the reflected one. Again, all of this is incorporated in the Standard Theory in a natural way.

After these results were understood, it was realized that the combined product operation CP was, remarkably, a symmetry of the weak interactions as well as the strong and electromagnetic ones—the combined effect of both P and C on an equation describing a system always led to a new equation that described a realizable system. These insights had a significant effect on helping to build the Standard Theory.

Then in 1964 an experiment found that the weak interactions were not perfectly CP symmetric, but this time the violation was tiny instead of maximal. One way to present the size of the effect is as follows. Let A represent the probability for a particular meson, called $K_L$, to decay to a negatively charged pion ($\pi^-$), plus a positron, plus an electron neutrino, and A′ represent the probability for the same particle to decay to a positively charged pion ($\pi^+$) plus an electron, plus an electron antineutrino. A and A′ are examples of processes related by a CP transformation—$K_L$ is its own antiparticle, and the others are transformed into their antiparticles. If CP were a good symmetry, then $A = A'$. Experiment gives $A/A' = 1.0066$ to good accuracy. The ratio deviates from unity by only a few parts in a thousand.

For thirty years, experiments with decays of the K meson have been the only place where the effects of CP violation have been detected, in spite of considerable effort to find more. We now understand that mesons with a b-quark in them should also be a very good place to study effects of the CP operation. Because of that, both the United States (at SLAC) and Japan (at KEK) plan to construct colliders that produce large numbers of b-quarks (b-factories); these facilities could begin to take data as early as 1997. In addition, particles such as electrons and neutrons can have properties that show evidence of CP violation; it is very important to find out whether they do.

There are two particularly strong reasons why many researchers hope a better knowledge of the CP symmetry physics may be an important clue to deeper understanding. First, if three families of quarks and leptons exist, then it is automatically possible to incorporate a description of the observed CP violation into the Standard Theory—the resulting equations necessarily include an amount of CP violation approximately equal to what is observed. But if only one or two families exist this is not possible. The reason is very technical (two-by-two matrices relating electroweak and mass eigenstates of the quarks can always be written as real numbers, but three-by-three matrices necessarily have complex number entries). Here is a connection of an observed effect with the great puzzle of why there are three families! Until CP violation is better understood we will not know if this connection is in some sense related to why families exist, but it is exciting to hope for some relationship.

Second, one of the conditions stated by Sakharov (see Chapter 12) to be needed in order to explain the baryon asymmetry of the universe is that CP violation must exist. That it does exist is therefore very encouraging; but the situation is more complicated. The CP violation observed in the decays of kaons does not automatically enter the relevant equations in the way needed to produce the baryon asymmetry, and the amount of CP violation needed may be larger than what is observed. Models for the baryon asymmetry do exist in which the same CP violation accounts for everything, and other models in which it does not. New experimental data about CP violation is badly needed to constrain and focus the theoretical approaches.

# LIST OF SYMBOLS

| | |
|---|---|
| e | electron |
| $\mu$ | muon; Greek mu |
| $\tau$ | tau; Greek tau |
| $\nu_e$ | electron neutrino; Greek $\nu$ sub e |
| $\nu_\mu$ | muon neutrino; Greek $\nu$ sub mu |
| $\nu_\tau$ | tau neutrino; Greek $\nu$ sub tau |
| u | up quark |
| d | down quark |
| c | charm quark (or speed of light depending on context) |
| s | strange quark |
| t | top quark |
| b | bottom quark |
| W | W's are electroweak gauge bosons, with electric charge $+1$, $-1$ |
| Z | Z is an electroweak gauge boson with no electric charge |
| $\theta_W$ | electroweak unification angle; Greek theta sub w |
| $E = mc^2$ | Einstein's equation giving the amount of energy that can be obtained by converting mass into energy, or the mass of particles that can be created by converting energy from a collision into mass of created particles; c is the value of the speed of light. |
| $F = m \times a$ | Newton's second law |

| | |
|---|---|
| G | Newton's constant |
| h | Planck's constant |
| c | speed of light (or charm quark, depending on context) |
| $\gamma$ | photon; Greek gamma |
| $\Omega$ | ratio of amount of dark matter of a given kind in the universe to the amount just needed to eventually slow the expansion rate to zero; Greek capital omega |
| g | gluon |

$10^1 = 10$, $10^2 = 100$, $10^3 = 1{,}000$, $10^6$ = one million, $10^9$ = one billion, $10^{-1} = 1/10$, $10^{-2} = 1/100$, $10^{-6}$ = one millionth, $10^{-9}$ = one billionth

# LIST OF ABBREVIATIONS AND ACRONYMS

| | |
|---|---|
| eV | one electron volt, a unit of energy |
| MeV | million electron volts |
| GeV | billion electron volts |
| TeV | million million electron volts |
| | |
| Fermilab | Fermi National Accelerator Laboratory, Chicago, IL |
| SLAC | Stanford Linear Accelerator Center, Palo Alto, CA |
| BNL | Brookhaven National Laboratory, Long Island, NY |
| CERN | Centre Europeén Recherche Nucléaire, Geneva, Switzerland |
| DESY | Deutsches Elektronen-Synchrotron, Hamburg, Germany |
| CESR | Cornell Electron Synchrotron, Ithaca, NY |
| KEK | National Laboratory for High Energy Physics, Tsukuba, Japan |
| BEPC | Beijing Electron-Positron Collider, Beijing, China |
| LEP | Large Electron-Positron Collider, at CERN |
| PETRA | electron-positron collider at DESY, now closed |
| SPEAR | original electron-positron collider at SLAC |
| HERA | electron-proton collider at DESY |
| SLC | linear electron-positron collider at SLAC |
| LHC | planned large hadron collider at CERN, to take data about 2005 |
| NLC | the "next" linear collider—the possibility of proposing such a collider is under study by |

| | |
|---|---|
| | several countries, particularly the United States, Japan, and Germany—there are discussions about making it an international facility; it would be an electron-positron collider with energy four or more times that of the SLAC SLC |
| SSC | the Superconducting SuperCollider, now cancelled—see end of Chapter 3 |
| SLD | detector at SLC |
| CDF | detector at Fermilab |
| D0 | detector at Fermilab |
| ALEPH | detector at LEP |
| L3 | detector at LEP |
| DELPHI | detector at LEP |
| OPAL | detector at LEP |
| QED | quantum electrodynamics, the relativistic quantum theory of electromagnetic interactions |
| QCD | quantum chromodynamics, the relativistic quantum theory of strong interaction |

# GLOSSARY

accelerator

Accelerators are machines that use electric fields to accelerate electrically charged particles (electrons, protons, and their antiparticles) to higher energies. If accelerators are linear they need to be very long to achieve the desired energies, so most are circular and some use magnets to bend the particles around and back to the starting point, giving them a little extra energy each time around.

accelerator physics

The field of physics that studies how to accelerate particles to higher energies, and how to make accelerated beams of higher intensity (luminosity).

antiparticle

Every particle has an associated antiparticle, another particle with the same mass but with all charges opposite. If a particle has no charges, e.g., the photon, it is its own antiparticle. Often the antiparticle is denoted by writing a bar over the particle name, e.g., $\bar{e}$ for the electron antiparticle (also called the positron)

atom

An atom has a nucleus surrounded by electrons, bound together by the electromagnetic force. Ninety-two different atoms occur naturally, making ninety-two different chemical elements. The diameter of an atom is about 10,000 times larger than the diameter of its nucleus.

b-factory

A b-factory is a facility designed to produce and detect large numbers of b quarks, at least 100 million a year. Planned b-factories are electron-positron colliders, but a proton collider could also be used if an appropriate detector could be made. Their main goal is to study CP violation.

baryon

A baryon is a composite particle made of three quarks, any three of the six. Protons and neutrons are baryons.

baryon asymmetry

Our universe seems to be made of baryons, but not antibaryons, so there is an "asymmetry." Several ideas exist to explain how a universe could initially be symmetric, with equal numbers of protons and antiprotons, but evolve into our asymmetric one, with about a billion protons for every antiproton, today. This is an active research area.

beams

One way to learn more about particles is to cause them to collide with one another and see what happens. Beams of electrons and protons can be made by knocking apart hydrogen atoms and applying electric fields. Positrons and antiprotons don't exist naturally, because they annihilate as soon as they encounter an electron or proton; they can be made by hitting a target with energetic protons or electrons, and then collected by putting magnets behind the target, arranged so as to bend each kind of particle in a different path. Then bunches of them are accelerated to higher energies. When a particle hits a target, every kind of particle is made with a certain probability, so other beams of particles can also be made (neutrons, muons, kaons, neutrinos, etc.) by judicious arrangements of magnets and material.

big bang

Several strong kinds of evidence imply that our universe began as a tiny dense gas of particles that has been expanding since; i.e., our universe began in a "hot big bang."

boson

Bosons are any particles that carry an integer unit of spin (0, 1, . . .). They have different properties from particles with half a unit of spin (fermions). In particle physics "boson" has a more specific usage—bosons (photons, gluons, W's, and Z) are particles that are the quanta of the electromagnetic, strong, and weak fields. They transmit the effects of the forces between quarks, leptons, and themselves. Higgs bosons are quanta of a hypothetical Higgs field; they have not yet been detected.

charge—electric, color, weak

Each particle can carry several kinds of charge that determine how it interacts with others. Electric charge is familiar. Particles can have positive or negative electric charge, or none. Color charge and weak charge are not familiar because their effects can only be felt at distances smaller than the size of a nucleus. A particle cannot have random amounts of charge; only certain discrete amounts are allowed. The extent to which a particle feels each force is proportional to its associated charge. Quarks and gluons carry color charge; quarks and leptons, and W and Z bosons, carry weak charge.

chemical elements

Ninety-two different stable nuclei can be formed from neutrons and protons bound together. Each forms atoms by binding as many electrons to the nucleus as it has protons (so the nucleus is electrically neutral), giving ninety-two different atoms. These atoms are the smallest recognizable units of the ninety-two chemical elements.

cold dark matter

Particle physics theories that extend the Standard Theory often predict the existence of new, stable particles that were present in the early universe and survive to the present, making up a large fraction of the matter of the universe. These particles interact weakly and are usually massive, so they move slowly—they are cold. A possible example of such a particle is the lightest supersymmetric partner. Astronomers have evidence from galaxy motion, and from the large-scale structure of the universe, that cold dark matter exists.

color charge

*See* charge.

color field

Any particle carrying color charge has an associated color field around it. Any other particle carrying color charge feels that field and interacts with the first particle.

color force

The force between two particles carrying color charge. The color force (or strong force) binds quarks into protons and neutrons. The residual color force outside protons and neutrons is the nuclear force that binds protons and neutrons into nuclei. The color force is mediated by the exchange of gluons.

composite

Any object made of other objects is composite, as are atoms, nuclei, and protons. If quarks and leptons had followed the historical trend that each level of matter turned out to be composites made of smaller constituents, experiments should already have shown evidence of their compositeness. That, combined with theoretical arguments, suggests quarks and leptons may be the ultimate constituents of matter, the indivisible "atoms" of the Greeks.

constituents

Any objects that are bound together to make larger objects. *See* composite. For example, atoms are constituents of molecules, nuclei are constituents of atoms, etc.

cosmic rays

Protons and some nuclei that are ejected from stars, especially supernova explosions, move throughout all space. They impinge on the earth from all directions, and are called "cosmic rays." They normally collide with nuclei of atoms in the atmosphere, producing more "secondary" particles, mainly electrons, muons, pions, etc. A number of cosmic ray particles go through each of us every second, and they can interact in detectors and mimic interesting signals, so it is necessary to shield against them or be able to recognize them so they can be discounted as signals of new physics.

CP violation

Interactions of quarks, leptons, and bosons are normally invariant under a symmetry operation called "CP," the combined operations of "Parity" and "Charge Conjugation." A small violation of this invariance is observed, which may have important implications. *See* Appendix C.

dark matter

Particle physics theories that extend the Standard Theory predict several forms of matter that may exist in large quantities today throughout the universe, and make up most of the matter of the universe. Some move slowly and are called "cold dark matter"; others move rapidly and are "hot dark matter." Study of the motions of galaxies and of the formation of clusters of galaxies suggests that such dark matter exists, and theoretical cosmological arguments also suggests the dark matter exists. *See* cold dark matter; hot dark matter.

decay

The quarks and leptons and bosons that are the particles of the Standard Theory have interactions that allow them to make transitions into one another. Whenever one of them can make transitions into lighter ones, that transition will occur with a certain probability, and we say the heavier one is unstable and has decayed into the lighter ones. In the Standard Theory the up quark, the electron, and the neutrinos do not decay; the other fermions, and the W and Z, do decay.

Descriptive Understanding

An understanding of nature that describes how things work,

and which can explain experimental data for particle interactions, though not necessarily particle properties, is called a Descriptive Understanding in Chapter 7.

detector
: The properties of particles and their interactions are studied by observing their interactions and decays. These observations are done with detectors, which can be thought of as cameras that record information in several ways, not only on film. In order to learn about interactions at the forefront of today's questions, detectors have to be very large and use (and also often develop) better technologies. Every particle physics experiment has one or more detectors.

deuteron
: A deuteron is the second lightest nucleus (after the lightest, hydrogen, which has a single proton), composed of a neutron and a proton bound together by the nuclear force. The deuterium atom has a single electron bound to the deuteron (one electron because there is one proton).

Dirac equation
: The Dirac equation incorporates the requirements of both quantum theory and special relativity in the description of the behavior of fermions. It requires fermions to have the property called spin, and it predicts the existence of antiparticles. It was written by Paul Dirac in 1928.

electric charge
: *See* charge.

electron
: A fundamental particle. See Table 4.1 (also inside back cover) for its properties.

electron collider (short for electron-positron collider)
: One important way to study particle interactions and to search for new particles is to accelerate an electron and a positron to high energies and then collide them, using a detector to study what emerges. The energy to which they are accelerated is chosen to fit the question of interest. For

example, to study CP violation in b decays the energy is chosen to maximize the production of b quarks in an appropriate way, while to produce new heavy particles the energy is made as large as possible. All uses of electron colliders require very large luminosity (intensity).

electroweak force

The descriptions of the electromagnetic and weak forces have been unified into a single description, the electroweak force. The electromagnetic and weak forces appear different because the W and Z bosons that mediate the weak force are massive, while the photon that mediates the electromagnetic force has no mass; the electroweak unified theoretical description treats all the bosons on an equal footing.

electroweak unification angle

The electroweak unification angle, $\theta_W$, arises in the unified description of the weak and electromagnetic forces, basically in defining which boson is the photon. One of the reasons many particle physicists believe that the Standard Theory will be extended is that $\theta_W$ is a parameter that has to be measured in the Standard Theory, but in some grand unified supersymmetric theories it can be predicted.

family

Quarks, and leptons, come in three families (Fig. 4.1). The universe we see is made up of the first family (the up and down quarks, the electron, and its neutrino). The other families differ only in that their particles are heavier. We do not yet understand why there are three families.

fermion

Fermions are particles with half a unit of spin. They have different properties from particles with an integer unit of spin (bosons). Quarks and leptons, the matter particles, are fermions.

Feynman diagrams

The rules of any quantum field theory can be formulated so that it is possible to draw a set of diagrams representing the processes that can occur, and to assign a probability of occurrence to the process represented by each diagram. *See* Appendix A.

field

Every particle is the origin of a number of fields, one for each nonzero charge the particle carries. Interactions occur when another particle feels the field of the first one (and vice versa). There are electromagnetic fields, weak fields, and color (strong) fields. Any particle with energy (including mass) sets up a gravitational field. In the Standard Theory particles get mass by interacting with a Higgs field, but the origin of the Higgs field is not yet understood.

flat universe

The universe is said to be flat if its expansion rate is slowing just so that infinitely far in the future the expansion rate becomes zero. The expansion rate is determined by the total mass of the universe, because the self gravitational attraction causes the slowing. The observed amount of mass is not yet sufficiently well measured to be sure of the answer, but it is near the amount to make a flat universe. If the universe is flat, most of the matter in it must be dark matter, not made of quarks and electrons.

forbidden

Processes can be naively imagined that might occur, but should not occur according to the predictions of the Standard Theory. Whether they occur is then a test of the Standard Theory. If they occurred at the same rate as other processes the Standard Theory would be wrong; if they occur at much smaller rates, or do not occur at all, they are giving us a clue to how to extend the Standard Theory.

force

All the phenomena we know of in nature can be described by four forces: the gravitational, weak, electromagnetic, and strong forces. Although the weak and electromagnetic forces appear different to us, they can be described as unified into one force (electroweak) in a more basic way; there is evidence that a similar unification of that electroweak force with the strong force also occurs. The attempt to unify all four forces is an active research area. In particle physics, "force" and "interaction" mean essentially the same thing.

gauge boson
: The strong, electromagnetic, and weak interactions are transmitted by the exchange of particles called gauge bosons (gluons, photons, and W's and Z's). The gauge bosons are the quanta of the strong, electromagnetic and weak fields.

gauge theory
: A quantum field theory where interactions occur between particles carrying charges, with strengths proportional to the sizes of the charges, and are transmitted by bosons that are quanta of the fields set up by the charges, is called a gauge theory.

glueballs
: Hadrons can be formed of gluons alone, without any quarks. They are called glueballs.

gluino
: The hypothetical supersymmetric partner of the gluon, differing only in that the gluino has spin 1/2 while the gluon has spin 1, and the gluino is heavier.

gluon
: The particle that transmits the strong force, the quantum of the strong field.

gluon jet
: The color or strong force is so strong that colored particles (quarks and gluons) hit or produced in a collision can only separate from other particles carrying color charge by binding to one another, and making color-neutral hadrons. Thus an energetic gluon or quark becomes a narrow "jet" of hadrons as it moves along, turning its energy into the mass and motion of several hadrons. A quark or gluon appears in a detector as a jet of typically five to fifteen hadrons.

grand unification

The proposed unification of the weak, electromagnetic, and strong forces into a single force. This unification, if it occurs, must happen in the sense that the forces act as a single one at very short distances, a million billion times smaller than distances that have been studied experimentally so far; the forces behave differently when they are studied at larger distances.

graviton

The quantum of the gravitational field, that mediates the gravitational force. Since gravity plays essentially a spectator role for the Standard Theory, gravitons and their properties have not been discussed in this book.

hadron

The properties of the color force and the rules of quantum theory allow certain combinations of quarks (and antiquarks) and gluons to bind together to make a particle; all such particles are called hadrons. When it is mainly three quarks that bind, the resulting hadron is called a "baryon." When it is quark and antiquark it is called a "meson," and when it is gluons it is called a "glueball." Hadrons all have a diameter of about $10^{-13}$ cm. The proton and neutron are the most familiar baryons. Pions are the lightest mesons and thus are produced most frequently in collisions. Kaons are the next lightest hadrons and have properties that make them useful in many studies.

helium abundance

As the universe cools after the big bang, it eventually reaches a stage (about a minute after the beginning) when protons and neutrons form, and then nuclei. Nuclei up to helium can form, but collisions are too soft for heavier nuclei to form. The theory of the big bang allows the fraction of nuclei that are helium to be predicted, and that fraction can be measured; the observed amount agrees very well with the predicted amount. This is one of the main reasons why it is generally believed that the universe began in a hot big bang.

Higgs boson

The Higgs boson is the quantum of the Higgs field. It will be possible to produce and detect Higgs bosons at the CERN LEP collider if they are light enough, and at other upgraded or future colliders. *See* Higgs field.

Higgs field

*See* Higgs mechanism. In the Standard Theory particles (bosons and fermions) are thought to get mass by interacting with the Higgs field. The Higgs field must have very special properties for the masses to be included in the theory in a consistent way. The other fields we know of arise from particles that carry charges, but we do not understand how the Higgs field could arise; that is why physicists are very nervous about the Higgs physics, and not yet in any agreement about whether Higgs bosons exist.

Higgs mechanism

The Higgs mechanism is a special set of circumstances that must hold if bosons and fermions are to get masses from interacting with a Higgs field, even if the Higgs field exists. In the Standard Theory these circumstances can be imposed, and in some extensions of the Standard Theory they can be derived.

high-energy physics

Another name for particle physics, often used because much of particle physics is based on experiments requiring high-energy beams.

hot dark matter

*See* dark matter.

infinities

Some quantities that should be calculable in relativistic quantum field theories seem to have infinite values. In the past this was a major problem in understanding the theories.

Slowly over sixty years it was learned that a proper formulation of the theory does not have the infinities, and today it is known how to avoid their appearance. The procedure to formulate the theory so the infinities are not relevant is called renormalization.

inflationary universe

According to the inflationary universe theory, as the universe expands after the big bang, it goes through a stage of very rapid expansion, called "inflation," and then slows down to the present rate.

Input and Mechanism Understanding

To have more than a Descriptive Understanding, it is necessary to be able to calculate the values of quantities that are normally input into the theory (such as masses), and to understand such things as which solutions of equations actually describe our world; see Chapter 7.

interaction

*See* force.

jet

*See* gluon jet.

kaon

One of the hadrons. *See* hadron.

Lamb shift

The difference in energy between two closely spaced energy levels of an atom. The measurement of the Lamb shift for hydrogen in 1947 stimulated theorists to learn how to calculate it in spite of the infinities the theory seemed to produce for its value, and led to the successful implementation of the renormalization procedure.

lepton

A class of particles defined by certain properties: leptons are fermions with spin 1/2 that do not carry color charge, and that have another property called lepton number that is different for each family. The known leptons are the electron, the muon, the tau, and their associated neutrinos.

linear electron collider

Particles that travel in a curved path radiate photons that carry away the energy of the particle. This happens for lighter particles with greater probability than for heavier ones. For electrons this loss of energy is a large effect at the CERN LEP collider, so it is unlikely that any future collider will be circular. The next electron collider built is expected to be a linear one (NLC, for Next Linear Collider), modeled on the first linear collider, the SLC at SLAC.

luminosity

Any collider has two basic figures of merit, the maximum energy it can supply to the collisions, and how often it can cause collisions to occur. The number of events at a collider over some period of time is the product of two factors, the probability that if two particles actually collide something will happen, and the number of collisions. The latter is just a property of the collider, not of the physics that governs the collision. It is called the luminosity. It depends on features such as how many particles can be accelerated, how tightly bunches of them can be packed, and so forth.

mass

Mass is an intrinsic property of any object, measuring how hard it is to make the object move. It can be thought of as weight, though the two are not quite the same; the mass of an object does not change, but if it were transported to a different planet its weight would change.

matter

It is useful to think of quarks and leptons as the basic particles that make up all the things around us, and the photons and gluons that bind them as quanta of the fields. We call the quarks and leptons matter particles. Sometimes "matter particles" is used to mean fermions.

Maxwell's equations

Electromagnetism, the unified theory of all electric and magnetic phenomena, is summarized in a set of equations, first written by James Clerk Maxwell in the 1860s. When they are extended to include the effects of the quantum theory the theory of quantum electrodynamics (QED) is obtained. Physics students spend about a quarter of their time for two years learning how to solve Maxwell's equations, unless they plan to work in a subfield that relies heavily on Maxwell's equations, in which case they spend much more time studying them.

mediate

The effects of interactions are transmitted from one particle to another by exchange of particles called bosons. The bosons are said to mediate the interaction or force.

meson

*See* hadron

microwave background radiation

As the universe expanded and cooled, the original particles decayed or annihilated until only photons, neutrinos, and protons, neutrons, and electrons that formed atoms were left. Today there is a cold gas of photons, about 400 in each cubic centimeter of the universe, called the microwave background radiation, because the wavelength of the photons is in the microwave part of the spectrum.

molecules

Although atoms are electrically neutral, the positive and negative charges are not at the same place so there is some electric field outside an atom (*see* Figure 4.2). Therefore atoms can attract one other, and form molecules, which can get very large.

muon

A fundamental particle. *See* Table 4.1 (also inside back cover) for its properties. A muon decays into an electron and neutrinos in about one-millionth of a second. Muons are made in collisions at accelerators, and in decays of other particles produced at accelerators and in cosmic ray collisions.

national laboratories

Since much of the research in particle physics has to be done at large accelerators that are very expensive, the accelerators are built as national facilities at a few labs and used by all particle physicists. Chapter 5 describes these facilities.

neutrino

A fundamental particle. See Table 4.1 (also inside back cover) for its properties. There is one neutrino for each of the three families.

neutron

*See* hadron. Free neutrons decay into a proton and an electron and an antineutrino with a lifetime of about fifteen minutes; when the neutrons are bound into nuclei (such as those in us) the decays are no longer possible because of subtle effects explained by quantum theory, so the neutrons in nuclei are as stable as protons.

Newton's constant G

Newton's law of gravitation says that the gravitational force between any two bodies is proportional to the product of their masses and decreases as the square of the distance between them. This statement is turned into an equation by inserting the constant G, so the force $F = Gmm'/r^2$.

Newton's laws

Newton wrote the law that describes the gravitational force (*see* Newton's constant G), and three laws that describe motion. The first law says that every moving body moves in a straight line at constant speed unless a force acts on it, the second law says that the product of the mass of a body and its acceleration is equal to the force acting on it—$F = ma$, and the third law says that if one body applies a force on a second then the second applies an equal and oppositely directed force on the first.

nucleus

Although protons and neutrons are color-neutral composites

of quarks and gluons, the quarks and gluons are not all at the same places so some of their color fields exist outside the proton or neutron, giving an attractive force that binds protons and neutrons into nuclei. The attractive effects of this residual color force are offset by the electrical repulsion of the protons, so nuclei with too many protons cannot exist. It turns out that there are ninety-two stable or long-lived nuclei in nature. They are the nuclei of the atoms of the ninety-two chemical elements.

particle

"Particle" is used somewhat loosely, and includes not only the elementary quarks and leptons and bosons, but also the composite hadrons. It also includes any (currently hypothetical) new particles that might be discovered, such as the supersymmetric partners of the quarks and leptons and bosons.

particle physics

This is the field of physics that studies the particles and tries to understand their behavior and properties. Sometimes a distinction is made between particle physics that studies quarks, leptons, gauge bosons, and Higgs physics, and the study of hadron physics that aims to relate the properties of the hadrons to the theory of the color force.

photino

The (hypothetical) supersymmetric partner of the photon.

photon

The photon is the particle that makes up light. It transmits the electromagnetic force. It is the gauge boson of electromagnetism.

pion

The lightest hadron, and therefore the one most often produced in collisions.

Planck's constant h

In the quantum theory many things are quantized, such as the energy levels of atoms. h sets the scale of quantization—

energy levels are separated by amounts proportional to h, the amount of spin a particle can have is a multiple of h, etc.

Planck scale

We measure all things in units. The three quantities Newton's constant G, Planck's constant h, and the speed of light c, can be combined in various ways to make all possible units. The natural way for humans to choose units is to use units such as centimeters, seconds, etc. that are small but that we have experience with. If one asks what are the natural units for a universe ignoring the humans, then presumably it is the units formed from G, h, and c; they give us the natural scales of the universe itself. For time the resulting unit is about $10^{-43}$ seconds, and for distance it is about $10^{-34}$ centimeters. These are called the Planck time and the Planck distance, because their relevance was first discussed by Max Planck.

pointlike

If matter is probed with projectiles that are large and that have energies less than what is needed to change the energy levels of an atom, then atoms will seem to be pointlike objects. If the energy is increased, eventually the projectile will penetrate the atom but encounter the nucleus, which will seem to be pointlike. With higher energy the nucleus will appear to be made of pointlike protons and neutrons. With still higher energies the protons and neutrons will be seen to be made of pointlike quarks and gluons. As the energies of projectiles were increased still more, quarks and leptons might have been seen to be made of something still smaller, but that has not happened—they behave as pointlike up to the highest energies they have been probed with, energies well beyond those for which we would have expected to find more constituents if history were to repeat itself once more.

positron

The antiparticle of the electron.

predict

The verb "predict" is used in the normal sense that a theory may predict some unanticipated or as yet unmeasured result. It is also used in another sense: a theory can be said to predict a result that is already known, because once the theory is written it gives a unique statement about that result. Sometimes an in-between situation holds, in that the theory predicts a result uniquely in principle, but the prediction depends on knowing some other quantity or requires very difficult calculations.

primary theory

The name used in this book for the theory sought by many particle physicists that not only includes the Standard Theory but also includes the theory of gravity, and explains why the theory takes the form it does, and explains what quarks and other particles are, and explains what space and time are, and more. Knowing the primary theory would give us a Why Understanding. *See* theory of everything.

projectile

To study particles and their interactions it is necessary to probe them with projectiles. The projectiles are other particles, electrons, and photons and neutrinos and protons because these are the only things that are small enough and can be given enough energy to help us learn more about the particles.

proton

*See* hadron.

proton decay

If the Standard Theory were the complete theory that described nature, protons would be stable, never decaying. If the Standard Theory is part of a more comprehensive theory that unifies quarks and leptons, then probably protons are unstable, with extremely long lifetimes. Experiments that search for proton decay are very important, because if we knew it occurred (and what the proton decayed into), those facts would provide valuable information about how to extend the Standard Theory.

quantum

Each particle is surrounded by a field for each of the kinds of charges it carries, such as an electromagnetic field if it has electrical charge. In the quantum theory the field is described as made up of bosons that are the quanta of the field. More loosely, the smallest amount of something that can exist.

quantum field theory

When interactions among particles are described as transmitted by exchange of bosons the methods of quantum field theory are used.

quantum theory

The quantum theory provides the rules to calculate how matter behaves, at every level. Once scientists specify what system they want to describe, and what the interactions among the particles of the system are, then the equations of the quantum theory are solved to learn the properties of the system.

quark

A fundamental particle. See Table 4.1 (also inside back cover) for a list of quarks and their properties.

quark jet

Because quarks must end up in hadrons, quarks that emerge from collisions or are produced actually appear in detectors as a narrow jet of hadrons, mostly pions. *See* gluon jet.

radioactive decay

Some nuclei are unstable, but live long enough to exist as matter until they decay. When they decay, they can emit several particles, photons, electrons, positrons, neutrinos, neutrons, and even helium nuclei. For historical reasons such decays were called "radioactive decays." Sometimes scientists use the emitted particles as tools to do experiments.

reductionism

One way to study the natural universe is to study very detailed aspects of nature, to take things apart and see what they are made of, and to focus on small steps. This approach is called "reductionist." It has been a powerful success, letting us build up the remarkably complete picture of nature we now have. Whenever possible scientists have tried to unify what was known. Recently in particle physics the trend toward unification has been increasingly successful.

relativity

Whenever particles can move at speeds near the speed of light (called relativistic speeds), and whenever fields are involved, the description of nature must satisfy the requirements of Einstein's "special relativity" theory.

rules

In order to have a complete understanding of nature, it is necessary to know the particles, the forces that determine the interactions of the particles, and the rules for calculating how the particles behave. For the motion of objects normally on earth or in the sky the rule to calculate is Newton's second law, $F = ma$. When atomic distances or smaller are involved, the Schrödinger equation replaces Newton's second law. In particle physics additional relativistic requirements are added to make the complete set of rules.

Schrödinger equation

The equation from quantum theory that tells how to calculate the effects of the forces on the particles. It is the quantum theory equivalent of Newton's second law.

science

Science can be defined as a self-correcting way to get knowledge about the natural universe, plus the body of knowledge obtained that way. It is both a method and the resulting understanding and knowledge. The method requires making models to explain phenomena, testing them experimentally, and revising them until they work. The goal of science is understanding. Once part of the natural world is understood, it may be possible to develop applications of the new knowledge. The process of developing such applications is properly called technology, not science. Although scientific knowledge may, and usually does, lead to technology, science is not necessary for technology, and technological developments have led to new science as much as the opposite. Before the time of Galileo many technological developments occurred that had no scientific connection. From the time of Maxwell and his writing of the electromagnetic theory almost all technological developments have depended on earlier science. In recent years the words "science" and "technology" have been frequently misused, as if they were interchangeable.

selectron

The supersymmetric partner of the electron.

signature

A new particle will have some characteristic behavior in a detector that allows it to be recognized. All particles will decay into others in a unique way that is different for every particle. Knowing the properties of the particle allows us to calculate how it will decay. The features that allow a new particle to be identified in a detector are called its signature.

slepton

A superpartner of any of the leptons.

solar neutrinos

The reactions that fuel the sun lead to the emission of photons that reach the earth as sunlight, and of neutrinos that we do not see with our eyes, but which can be detected in special neutrino detectors. At present there is great interest in these neutrinos because the number being detected is fewer than expected, and this may be a signal that neutrinos have mass, in which case we could account for the decrease in their number. If they have mass, the experiments to detect them will allow the value of their mass to be measured.

sparticle

Any superpartner.

special relativity

The constraints of special relativity are two conditions that Einstein pointed out should be satisfied by any acceptable physical theory. Somewhat oversimplified, these conditions are, first, that no particle can move faster than the speed of light in vacuum, and second that scientists working in different labs moving with different relative speeds should formulate the same natural laws. The constraints imposed by these conditions have surprising implications for the structure of acceptable theories. For example, the Schrödinger equation of quantum theory does not satisfy these conditions. But when it was generalized by Dirac to do so, the resulting equation led to the prediction of antiparticles, which need not have existed from the point of view of quantum theory alone.

spectra

Atoms can exist in a number of discrete energy levels. They emit or absorb photons when they make transitions from one level to another. The energies of the photons emitted or absorbed by one atom are different from those of all other atoms. The photon energies are directly related to their frequencies, which set their colors, so by observing the colors of the photons it is possible to determine which atoms are being observed. This can be done in a laboratory, and it can also be done with the light reaching us from stars, near or distant, which allows us to identify the atoms that stars are made of. Only the same ninety-two elements we find on earth are seen throughout the universe.

speed of light

Light and all other massless particles travel in vacuum with a speed, usually labeled c, whose value is about 300 million meters a second. Special relativity implies that no particle or signal can move faster than the speed of light, and that photons always have this speed regardless of the speed of their source.

spin

Spin is a property that all particles have. It is as if particles were always spinning at a fixed rate (which could be zero), which can be different according to the type of particle, but not quite the same because the particles do not have to have spatial extension to have spin. The amount of spin is required by the quantum theory to come in definite amounts; if the unit is chosen to be Planck's constant h divided by $2\pi$ then particles can have zero spin, half a unit of spin, one unit of spin, etc.

spontaneous symmetry breaking

Often the equations of a theory may have certain symmetries, but their solutions may not. For example, the equations may describe several particles in identical ways, but the solutions may give the particles different properties. Some examples are given in the book, particularly in Chapter 7. When this occurs under certain conditions it is called spontaneous symmetry breaking.

squark

The superpartner of any of the quarks.

Standard Model

The very successful theory of quarks and leptons and their interactions that is described in this book is called the "Standard Model" by particle physicists. The name arises historically as the theory develops, and then is difficult to change because it is widely used. Because the Standard Model is the most complete mathematical theory ever developed, and is well tested experimentally, I have called it the "Standard Theory" in this book.

Standard Theory

*See* Standard Model.

structure

Objects have structure if they have parts—if they are made of other things. Whether objects have structure can be learned from experiments that probe them with projectiles. Over the past century each stage of matter found as it became possible to search for smaller things turned out to have structure. Quarks and leptons appear not to have structure, so perhaps the search for the basic constituents has finally ended.

subatomic particle

Any particle that is contained in an atom, or any particle that can be created in collisions of such particles, is loosely called "subatomic," whether it is composite like a proton, or elementary like a quark or electron.

superpartner

If the theory that describes nature has a symmetry called supersymmetry, then every normal particle (the ones we know) has associated with it a partner that differs only in its spin and its mass.

superstring

Experiments suggest that quarks and leptons do not have size or structure, and today's theory represents them as pointlike sources of quantum fields. The superstring theories argue that particles should be represented as strings, probably closed loops, rather than points, but of a size much smaller than could be seen directly in experiments. Today's description would be a valid way to approximately describe particles at the distances they can be probed in experiments. Proponents of superstring theories claim that they have the potential to achieve what we have called a Why Understanding.

supersymmetry

A hypothetical symmetry of the theory that describes nature, which says that even though fermions and bosons seem to us to be very different in their properties and their roles, in the theory itself they appear in a symmetric way. If supersymmetry is indeed realized in nature, then every particle has a superpartner.

technology

*See* science.

theory

The word "theory" is usually used precisely in physics. Theories are sets of equations whose solutions describe physical systems and their behavior. Further description of the properties of a theory is given in Chapter 1.

theory of everything

A "theory of everything" would not only describe how things work, it would explain why things are the way they are. In this book our present understanding is called a Descriptive Understanding. A theory of everything would require in addition an Input and Mechanism Understanding and a Why Understanding. The name "theory of everything" is unfortunate in one way because it does not tell how to deduce the behavior of complex systems from a knowledge of their components. In this book I have used the name "primary theory" instead.

transmit

*See* mediate.

unification

Scientists have sought for centuries to unify the descriptions of apparently different phenomena by showing they were due to the same underlying natural laws, and that complex levels of matter were made of simpler levels. This unification process is a subject of very active research for the forces of nature today. The possible unification of the strong, electromagnetic, and weak forces is called a "grand unification." There is a continuing effort to unify these forces with gravity.

unstable

*See* decay.

vacuum

Any physical system will settle into the lowest energy state it can. For most fields that is the state where the field is zero, but theorists hypothesize that for the Higgs field a state where the field takes on a constant value different from zero gives the system a lower energy. The value of the Higgs field in that system is called its "vacuum expectation value."

vacuum expectation value

*See* vacuum.

weak charge

*See* charge.

weak force

*See* force. The weak force is described in Chapter 4.

wino

The supersymmetric partner of the W boson.

Why Understanding

Achieving a Why Understanding of nature requires not only having a complete description (Descriptive Understanding), and being able to derive the values of masses and other quantities needed for the description (Input Understanding), and being able to know which solutions of equations actually describe nature when there are alternatives (Mechanism Understanding), but also knowing why the forces are what they are, what quarks and leptons are, and more. Why Understanding is the goal of most particle physicists. *See* primary theory.

# INDEX